ROCKET AGE

The Race to the Moon and What It Took to Get There

George D. Morgan

Guilford, Connecticut

An imprint of The Rowman & Littlefield Publishing Group, Inc.
4501 Forbes Blvd., Ste. 200
Lanham, MD 20706
www.rowman.com

Distributed by NATIONAL BOOK NETWORK

British Library Cataloguing in Publication Information available

Library of Congress Cataloging-in-Publication Data Available

ISBN 978-1-63388-636-0 (hardback) | ISBN 978-1-63388-637-7 (electronic)

∞™ The paper used in this publication meets the minimum requirements of American National Standard for Information Sciences—Permanence of Paper for Printed Library Materials, ANSI/NISO Z39.48-1992.

To my parents, G. Richard Morgan and
Mary Sherman Morgan,
the greatest rocket scientists no one has ever known.

CONTENTS

1

A METHOD OF REACHING EXTREME ALTITUDES

The dream of yesterday is the hope of today, and the reality of tomorrow.

—Robert Hutchings Goddard

On October 18, 1899, a seventeen-year-old boy climbed a tree behind his family's barn and experienced an epiphany—a simple moment of thoughtful realization that would change the world. The boy's name was Robert Hutchings Goddard, and as he looked up into the clear blue Worcester, Massachusetts, sky, he became curious as to how one might ascend to the heights of the clouds, the Moon, and beyond. Of the experience he would later write, "On this day I climbed a tall cherry tree at the back of the barn . . . and as I looked toward the fields at the east, I imagined how wonderful it would be to make some device which had even the possibility of ascending to Mars, and how it would look on a small scale. I was a different boy when I descended the tree from when I ascended. Existence at last seemed very purposive."[1]

Goddard had already inherited an interest in science from his father—an interest that only accelerated when their home was wired for electricity and artificial light could be ordered up with the simple flick of a switch. Yet there was something different about the tree-climbing experience. It had such an impact on Goddard that for the rest of his life he quietly celebrated October 18 as the "Anniversary Day." His newfound interest in leaving Earth caused a slight change in his science interests, leaning him more toward technology, engineering, and physics. Unfortunately, his formal education was delayed due to a number of childhood illnesses, but in 1908 he graduated from Worcester

Polytechnic Institute, and did so as class valedictorian. Soon after graduation he became briefly engaged to a young woman who, fittingly, was the class salutatorian. The following year Goddard began his graduate studies at Clark University.

After his first year at Clark, Goddard began to contemplate the possibility of building a rocket using liquid propellants. Solid fuel rockets had several major drawbacks, not the least of which was the inability to regulate the speed at which they burned their propellants. Once ignited, a solid propellant rocket's thrust and burn rate remained essentially constant. Throttling of a rocket's engine would be necessary, Goddard theorized, in order to fly and control a rocket's trajectory over long distances and altitudes. On paper he began to calculate what would be needed to build such a missile.

Though he would eventually be credited with the invention of the liquid fuel rocket, in truth Goddard was reinventing a rocket theorized thirty years earlier by Konstantin Tsiolkovsky, a nineteenth-century Russian scientist. Tsiolkovsky's work had been hidden from much of the world due to a lack of enthusiasm for it in his native country. It was Tsiolkovsky who first theorized what would become the modern liquid fuel rocket design. He envisioned a multistage device that would contain tanks of liquid oxygen and liquid hydrogen, and a space at the top for a human to sit and control it all. He was also the first to calculate the minimum orbital velocity of an artificial satellite, five miles per second, and escape velocity, seven miles per second. His work would have been a great inspiration to Goddard, had he known about it, especially since Tsiolkovsky had backed up his theories with proven math and science. As a result, his work was little published in his own language, let alone English.

Another work-alone rocket recluse at this time was Hermann Oberth. Like von Braun and so many scientists then and since, Oberth was inspired by the writings of Jules Verne. Independent of both Goddard and Tsiolkovsky, Oberth began to develop theories of space flight—theories that were highly derided by his peers. He wrote a doctoral thesis based on those theories that was initially rejected as being "utopian."[2]

Even so, by the time Goddard was deep into his university studies he was able to obtain some of Tsiolkovsky's writings, the most significant of which involved the speed a man-made object would have to travel to break free of Earth, and how that might be achieved. In 1913 Goddard was working at Princeton University when he became ill again, this time with tuberculosis. He took a long leave of absence from his work, which afforded him the time to flesh out his ideas and theories for a rocket using liquid propellants. A year later he filed a patent for a liquid fuel rocket using gasoline and liquid nitrous oxide. This would turn out to be one of the most historic patents in rocket science.

In 1916 Goddard applied for a research grant from the Smithsonian Institution. In the application he included a short publication titled *A Method of Reaching Extreme Altitudes*. He was awarded $5,000—a tidy sum in 1916. Three years later the Smithsonian published Goddard's grant application paper; 1,750 copies were printed and sent around the world.

But Goddard would not be satisfied with mere "paper and pencil" work like Tsiolkovsky and Oberth; he wanted to build flightworthy rockets to test the many theories that were developing. He quietly ensconced himself in a small workspace in Worcester and began moving his designs from theory to blueprint to hardware. In November 1923 he test-fired the world's first liquid fuel rocket engine. Two and a half years later, on March 16, 1926, Goddard flew a small oxygen/gasoline liquid fuel rocket, launching it from a remote cabbage farm nearby Auburn. Its peak altitude: forty-one feet. An inauspicious beginning, yet that tiny homemade rocket with its humble, cork-pop apogee would one day change everything for everybody.

Once news leaked out about Goddard's liquid fuel rocket, Oberth decided to follow with some hardware experiments of his own. In the autumn of 1929 Oberth test-fired a small liquid fuel rocket engine. He was assisted in this project by an enthusiastic eighteen-year-old German protégé who had been following Oberth's work like religious gospel: Wernher von Braun. In the 1950s the teacher would become the student as Oberth found himself working under von Braun's direction in the U.S. Army's Redstone missile program.

As for Konstantin Tsiolkovsky, he would one day be lauded for his work, albeit posthumously. Among the many honors would be his statue in Moscow as one of the Soviet Union's "Conquerors of Space," a lunar crater named after him, and a Google Doodle.

2

BURNING DOWN THE HOUSE

It takes sixty-five thousand errors before you are qualified to make a rocket.

—Wernher von Braun

Ever since he was a little boy, the only thing Wernher von Braun wanted in life was to build a rocket and fly to the Moon. Many people have all-consuming passions that occupy their thoughts day and night, and for Wernher, this was his. In 1924, on Wernher's twelfth birthday, his father had given him a telescope. Like Goddard's tree-climbing experience, this simple event would change the course of history, for it would change the boy who owned it, who would make history. From that day on, Wernher spent many of his days reading Jules Verne, and his evenings peering into the heavens. By then Robert Goddard's treatise, *A Method of Reaching Extreme Altitudes*, had been translated into German, and Wernher could be found late at night reading it, and rereading it.

In his twenties, Wernher was given his first research grant to further German rocket science, and he was certain then that Germany would be the conduit to his lunar dream's fulfillment. For several years thereafter, von Braun had told his rocket enthusiast associates it would be Germany that would help them build their rocket to the Moon. But Wernher had grown since those idealistic days. After ten years of taking their money, he now realized that all military checks had only one end game—the propagation of weaponry. He and his fellow scientists were frustrated. In their hands lay a marvelous invention with tremendous peacetime potential—a potential that had been squandered and frittered away by madmen. Such misuse of human discovery troubled him, and always would.

Wernher von Braun examined the large wall map above him that indicated the positions of advancing Soviet and Allied troops. The war was lost—anyone who could read a map could see the obvious. But while some of his countrymen could see in such maps only a vision of the Third Reich government slipping away, von Braun saw something completely different: the destruction of his spacefaring dream.

A young soldier coughed to regain his attention, and von Braun resumed throwing his personal papers and documents into the young man's wheelbarrow.

The smoke could be seen for fifty miles—an unintended beacon for Soviet ground troops and Allied bombers looking for the notorious German rocket base at Peenemünde. Of course, this was war on a massive scale, and there were thousands of rising smoke columns throughout the country. Nothing unusual about a smoke column these days. Germany still controlled much of its airspace, but situated on the coast as it was, the base was especially vulnerable. A crooked finger of a peninsula twenty-five miles west of the border with Poland, Peenemünde had become a liability. The steady advance of the Red Army, combined with an increasing flow of Allied bombers, meant the facility's security could no longer be guaranteed.

A wooden pier extended two hundred meters into the ocean, its planks slowly rotting in the heavy saline air of the Balkan Sea. Approximately one hundred steel chemical drums were lined up along its downwind edge. The dependable North Atlantic breezes pushed westward, keeping the voluminous black plumes, which furiously poured from the drums, away from the nearby factory buildings. Scattered throughout the compound stood a contingent of Nazi *Schutzstaffel* (SS) officers, most of them carrying fully loaded machine guns. Peenemünde was technically a civilian operation, but in reality, the SS supervised all its activities. On this day, it was their responsibility to direct the destruction of a huge collection of secret documents and close the facility. SS Reichsführer Heinrich Himmler had ordered his men to keep their eyes on the large crowd of engineers, technicians, and forced laborers who delivered the never-ending stream of wheelbarrows loaded with blueprints, documents, invoices, and memos to their fiery end. As each worker arrived with his load, one of the SS officers would point to an available barrel, and the worker would begin tossing documents into it. As the worker turned to run back for more, the SS officer sprayed a shot of accelerant into the barrel to keep the process moving at a fast pace.

As this document destruction operation progressed, another set of technicians and workers loaded hundreds of tons of manufacturing equipment and rocket parts into the boxcars of a waiting train, its engine idling and at the ready to leave at any moment. Wernher von Braun's rocket research and construction facility would be no more—its land destined to return to the farms and cow pastures from which it was born.

Von Braun and his team had been ordered to move their technology to another site for safekeeping: Mittelwerk. A converted gypsum mine deep within the low-lying mountain of Kohnstein, Mittelwerk was spacious and deep—perfect for protecting Germany's highly advanced rocket technology from the now-daily Allied bombing raids. Mittelwerk had been opened for rocket manufacturing two years earlier, but only after being greatly enlarged by thousands of slave laborers from the nearby Mittelbau-Dora concentration camp.[1] Peenemünde had been kept open as an engineering and research center, but now everything would have to be moved to the safer underground facility. Von Braun tried to convince himself that this was an improvement, as everything would be centralized.

With his file cabinets now empty, and another wheelbarrow filled to overflowing, Wernher nodded to the soldier, who pushed the load toward an open factory doorway. At the same moment, an SS officer entered and made some new marks on the map, indicating that a Soviet battalion had advanced another forty miles in their direction. Von Braun was astonished at how fast Stalin's army was approaching. He and every German soldier and worker at Peenemünde had no misconceptions as to what would befall them if they failed to evacuate the facility before the arrival of the Russian soldiers. Some of the scientists would be spared, of course, in order for Stalin to extract from them Germany's advanced rocket technology. Everyone else would be shot on the spot—the notorious Red Army had little interest in taking prisoners.

Wernher von Braun put on his gloves, tightened his wool coat around him, and left the building. He was followed by two engineers who had become his closest associates and confidants—Dieter Huzel and Arthur Rudolph. Unknown to their superiors, the three men, along with a number of other top-level scientists, had already been whispering about an escape plan. It seemed apparent to everyone but a few Nazi diehards that the war was lost. Fighting a war on two fronts had been a major tactical error on Hitler's part—the combined forces of the Allies to the south and west, along with a large and devoted Red Army to the east, was far too formidable even for a war machine like Germany. And so, von Braun and his top engineers had been quietly considering their postwar chess moves. They would have to secret away from Nazi control their most crucial blueprints and designs. They would then need to smuggle those designs out of Mittelwerk and hide them in a place secure from everyone—Germans, Russians, the Allied armies. Finally, they would have to surrender to someone—but to which country? For von Braun the choice was clear—only the Americans had the financial capacity, engineering wherewithal, and poetic sense of adventure needed to help him fulfill his lunar-centric dreams.

That is, if the Americans didn't imprison or hang him first.

The men got into a waiting car, and its driver pulled away. There was a nervous moment at the facility gate, where two SS guards manned a checkpoint. For several days rumors were rampant that Himmler had secretly ordered the execution of key rocket scientists in order to guarantee that the technology locked in their heads would not fall into Soviet control if they were captured. But the SS guards merely checked the men's travel documents, then waved them through. Unbeknownst to von Braun or any of the other scientists, SS Commander Hans Kammler still had plans for them.[2]

As the car accelerated south, Wernher von Braun turned to take one last look at his beloved rocket facility. His dreams and plans for building a manned rocket to the Moon would have to wait for another day. If they came at all.

3

PT-109 AND THE BUTTERFLY EFFECT

Without PT-109 there never would have been a President John F. Kennedy.

—David Powers, White House Official,
Kennedy Administration

If Kennedy had not become President, we would never have had a Moon landing.

—Dieter Huzel, German aerospace engineer

On July 20, 1969, at precisely 20:18 Coordinated Universal Time (UTC), Neil Armstrong's left boot touched the surface of the Moon, one of the final phases in what is arguably the greatest achievement in the history of mankind: landing men on the Moon and returning them safely to Earth. Although history paints this achievement as the natural order of the progression of science, it was anything but.

In fact, the entire spectacular endeavor owes its existence to the Butterfly Effect.[1]

In the late evening of August 2, 1943, a convoy of four ships was returning from a supply run to restock a Japanese Army–held island. The moonless, pitch-black night combined with a deep mist had dropped visibility close to zero. If that were not dangerous enough, the waters through which the convoy traveled were lined with hazardous coral reefs. The ships plying this particular supply route were referred to unflatteringly by the Americans as the "Tokyo Express." What the Americans did not know was that the Japanese, weary of their vulnerable barges being routinely attacked and sunk by American Patrol Torpedo (PT) boats, had docked their barges and replaced them with a small fleet of well-armed naval destroyers.

The lead ship in the small fleet was the *Amagiri*, a two thousand ton Imperial Japanese warship. The *Amagiri*, along with its three sister ships, were racing through South Pacific waters at thirty-four knots—a very high rate of speed for such large vessels. Normally they would not be moving so fast, but the *Amagiri*'s captain, Lieutenant Commander Kohei Hanami, suspected that a gauntlet of torpedo-laden American boats and fighter planes were waiting for them somewhere ahead. His plan was to hurry past the enemy positions before they had a chance to plan and mount an attack. And so he had ordered his helmsman to run at near-top speed back toward the safety of their main port. By the time the sun rose above the Pacific horizon, he and his men would be safe.

Captain Hanami's concerns were not misplaced. At that same moment, a platoon of much smaller U.S. torpedo craft were lying in wait not far ahead. Each PT boat held an average-size crew of twelve—much smaller than the massive two-hundred-men Japanese destroyers. Though the route and schedule of the Tokyo Express was not totally predictable, based on intelligence reports and prior experience the American PT crews had a general idea of which nights and what hours to expect the barges.

One of those PT boats would be forever etched into American combat history and lore: PT-109, skippered by young, athletic, future U.S. president John F. Kennedy. Lieutenant Kennedy had attained his status as a PT boat captain through a combination of hard work, excellent grades, lifelong boating skills, and behind-the-scenes political machinations. From the outset of World War II, Kennedy had wanted to get into combat in the South Pacific, and now he had achieved that desire.

On the night of August 2, the new PT commander was concerned about attacking the Japanese under the given conditions. Everything seemed to be against them. No Moon. A misty fog. No visibility. And now, no radar. The convoy had only one PT boat equipped with radar, and it had been ordered back to port earlier in the evening. The PT boats lying in wait were as blind as could be, setting up what was about to become a perfect storm of chaos.

It is not known for certain who spotted whom first, but it appears the Japanese lookouts spotted the smaller PT boats at about the same moment that 109's forward gunner, Harold Marney, shouted, "Ship at two o'clock!" On the *Amagiri*, the captain, having been notified of the PT boat's location, had ordered his helmsman, "Ten degrees turn—full speed ahead!" intending to ram the much smaller craft. This was a highly unorthodox method of attack, and risked damage to the larger ship as well, but the complete lack of visibility had brought the two vessels into such close proximity that there was no time for either combatant to maneuver and mount any form of standardized attack or defense. So quickly did events unfold that the only PT crewman below decks, Motor Machinist's Mate "Pappy" McMahon, did not hear the warning shouts from above. According to McMahon, "I thought we hit a rock."[2] Blown out of the severed boat and sucked under the black water, McMahon miraculously surfaced several hundred feet from the burning wreckage.

Twelve men, including Kennedy, survived the collision. Two crewmen were killed. Kennedy and eight of the survivors were in waters teeming with shark and barracuda, and were scattered so far apart that most of them could neither see nor hear the others. Those that survived now had a new peril—the ocean was on fire. The PT boat engines utilized high-octane aviation fuel, which burned at temperatures above 160° Fahrenheit. The fuel tanks had been blown apart by the collision and fuel was spreading over a wide area of the water's surface. And it had ignited. Kennedy and his men were at risk of burning to death, and in fact several had already sustained second- and third-degree burns.

Kennedy swam to the wreckage, where he found the remaining two survivors, Maguire and Mauer, who were clinging to the now-separated bow. Fearing an explosion of the fuel still inside the forward tanks, Kennedy ordered the two men to jump into the water and swim a safe distance away. But soon the flames diminished, the risk of explosion subsided, and Kennedy ordered the men back to the bow.

The time was approximately 2:00 a.m.

It took several hours, but one by one Kennedy managed to find and round up every one of the survivors, who held on to whatever piece of the slowly sinking bow they could get a hand around. There was room for a few of the men to climb onto the wreckage, and those with open wounds, whose exposure to saltwater caused extreme pain, took those positions. The rest remained in the water.

At this point their situation was as dire as could be imagined. According to William Doyle, "They had no food, no water, no radio, no life raft, no medical supplies, and, it seemed, no way of getting to safety."[3] It was a pitch-black moonless night, so they had no vision, and owing to the current, they had only a vague idea of their location—a location which shifted moment by moment. As they floated in the water, the men began to talk about a rescue. What they did not yet know—owing to radio silence—was that no other navy crews were aware of their predicament, and the other PT boats had left the area.

When the sun rose that morning, the crew's already impossible situation worsened. The bow, to which they clung like a life raft, was now in full view to any passing Japanese ships, or to the deadly enemy pontoon planes, or to Japanese forces that inhabited most of the nearby islands. A few bursts from a well-positioned machine gun, and PT-109's crew would perish.

Kennedy decided that remaining with the boat's wreckage was too dangerous. He picked out an island he was pretty certain was too small for the Japanese to put a settlement on, and ordered his men to start swimming. McMahon, the only one injured too severely to swim, demanded to be left behind, lest he endanger the others by slowing them down. But Kennedy would have none of it. Grabbing a strap of McMahon's life vest in his teeth, Kennedy began swimming, towing McMahon behind him. The other men followed.

From the moment that decision was made, it took four hours of nonstop swimming to reach the lonely island of Plum Pudding—a tiny dot of nothing in the middle of the Pacific nowhere. By the time the men tore their bodies up on the sharp coral, and plopped onto the island's sandy beach, they had been in the water for a total of sixteen hours. Despite towing McMahon's dead weight, Kennedy arrived on the beach first.

Exhausted beyond their limits, the men could not even stand and were forced to crawl on hands and knees to reconnoiter their group. There was one other island in view, the large Japanese-held island of Kolombangara—heavily armed and, without a doubt, filled with binocular-toting Japanese soldiers. Lacking physical energy to move beyond a crawl, and exposed on the sand in daylight, the men were now at their most vulnerable. Plum Pudding appeared to be deserted. But just as the men congratulated themselves on picking an island too small to be of consequence to the enemy, a Japanese barge motored by. The men crawled behind some bushes and managed to avoid being spotted.

After resting, Kennedy and his men explored the island and quickly came to a realization—their picture-perfect South Pacific isle was in fact a death trap. There was no fresh water, no animals large enough worth consuming, the plant life was inedible, and the fish just offshore were uncatchable. Some of the trees had coconuts, but after eating them the men became sick. Their skipper made a quick survey of their individual talents and discovered that none of the men had ever received any sort of survival training.

Kennedy was smart enough to know he could not wait. The longer they postponed taking some sort of rescue action, the weaker their bodies would become. With every day that passed, their health would steadily decline due to dehydration and malnutrition.

The men slept in the open that night. In the morning, as the sun rose on the Pacific, Kennedy began to formulate a daring plan. After sundown he would take the signal lantern they had salvaged from their vessel, swim out into the channel, and attempt to signal any PT boats that happened to be passing by. When he described his plan to the men, they vociferously attempted to dissuade him. To the crew it seemed like a suicide mission, but to Kennedy it was their only option.

Kennedy's midnight swimming rescue attempt yielded only one result: a thoroughly exhausted commanding officer. When Kennedy returned to the island the next morning, he vomited, then fell asleep. Over the next several days the castaways would swim back and forth to several neighboring islands, never once spotting a friendly vessel. Each time they would expend valuable energy that they could not spare. Then one day two scouts—native Solomons working for the Americans—showed up in a canoe on the island where most of the PT-109 crew was staying as they waited for Kennedy to return from yet another rescue rendezvous attempt. The natives agreed to take a message back to the U.S. Navy. Unfortunately, no one had any paper. Soon after Kennedy returned, they suggested a message be scratched inside the husk of a coconut. Since the writable area was very small, the message was brief and cryptic:

NAURO ISL
NATIVE KNOWS POS'IT
HE CAN PILOT
11 ALIVE
NEED SMALL BOAT
KENNEDY

The native scouts paddled all night to reach an American-held post. The first navy personnel and officers they encountered refused to believe the scouts' astonishing tale. They and their coconut, however, were sent up the chain of command and eventually ended up in front of base commander Thomas Warfield. He decided to take a leap of faith and sent PT-157, skippered by Lieutenant "Bud" Liebenow, on a rescue mission. Hours later, as PT-157 rendezvoused at a predetermined point with Kennedy, PT-109's captain had only one thing to say.

"Where the hell you been?"

Three years later, at the urging of his wealthy and politically connected father, John F. Kennedy ran for U.S. Congress in Massachusetts' 11th Congressional district, and won.

4

DESPERATION, SALVATION, AND THE TRANSFER OF TECHNOLOGY

I was accused of sabotaging the war effort by having my heart set on space rather than destroying London.

—Wernher von Braun

Wernher von Braun was a wanted man. He was being hunted by both the Americans and the Soviets. As Germany's defeat approached, both nations had in their control captured generals and other officers of the Third Reich. The rooms of the German military detainees had been bugged for months, and many of the men talked far too freely. The security teams who monitored their conversations reported to their superiors that one name in particular kept coming up again and again: Wernher von Braun.[1] The imprisoned German generals often discussed with arrogant bravado how they would soon obtain their freedom as von Braun's rocket program crushed their enemies. Von Braun's name and status were known to the Allies as early as 1943,[2] and as their armies homed in on Berlin and other geographical targets, von Braun was high on the "most wanted" list of Germans they were looking for.

Wernher certainly must have been aware that the invading armies were looking for him. His own ego had made his notoriety certain—he never passed up an opportunity to get attention for his rocket program, and himself. That notoriety was an asset during times when research and production budgets needed to be expanded—not as advantageous if one is on the losing side of a war.

As he stood on the terrace of the luxury ski resort, Haus Ingeburg, Wernher gazed out at the snow-capped peaks and thousands of square miles of alpine forest surrounding him. He searched for approaching armies but saw none.

German intelligence had informed him that the French army was west of them, the Americans to the south. But on this crisp blue spring day, the only thing visible from his position was nature at its finest.

Wernher felt a twinge of pain and shifted his position. His left arm had recently been severely broken in an auto accident, and a surgeon had been forced to reset the bone twice. His arm was wrapped in a thick cast and propped up at a steep angle such that his forearm was level with his shoulder. The pain medication was wearing off, and Wernher tapped another pill into his mouth. A server brought a bottle of wine and a crystal glass on a platter and set them down on the stone balustrade.

"Danke."

The server poured a small amount of wine into the glass, then left. Of this day Wernher would later joke to a crowd of American journalists, "I was in an especially upbeat mood. Hitler was dead, and the hotel service was excellent."[3] Wernher, his brother Magnus, Colonel Dornberger, and a small number of friends and associates had been allowed to move to this remote resort for their personal safety. The SS guards that accompanied them were a deep concern, owing to rumors the German military wanted the scientists assassinated in order to keep their secrets secret. Even if the rumors proved untrue, if captured by the Soviets, they would be tortured for those same secrets, then executed. Only the Americans offered some possibility of safety, but in wartime everything is fluid and uncertain, and the German engineers had no way of knowing if they would be shot while attempting to surrender, even to the Americans.

At this moment, though, Wernher's thoughts were not on surrender, but on his close friend and fellow engineer, Dieter Huzel. Somewhere beyond the evergreen forests below, Dieter was leading a team of trusted friends and engineers on a mission to save a remnant of the technology they had spent the better part of their lives creating: the design and construction of high-power liquid fuel rockets. Equipped with only three trucks and a handful of forged documents, Dieter and his team had volunteered to take plans, blueprints, manuals, and rocket parts into the Harz Mountains and quietly conceal their cargo in an old mine, which would then be dynamited shut. The goal was to prevent the technology from falling into the wrong hands—and by "wrong hands," they meant any nation or army other than the one that would give shelter and protection to the surrendering German engineers. The technology, therefore, became a poker chip in the impending surrender negotiations, which they knew must come.

Colonel Dornberger, the man who had recognized von Braun's genius before anyone else, arrived on the terrace and approached his friend.

"Sie sind zuruck."

Wernher smiled, realizing his colleagues were safe. After being informed that Dieter's mission had been a success, the two men decided it was time to put their plan into action. A few days earlier they had received the news of Hitler's death. The next morning, they had awoken to discover their SS guards had vanished sometime during the night. Now a decision had to be made.

Surrendering to the Soviets would be suicide. And even though the French army was only a few miles away, the Nazis had committed atrocities during their invasion of France, and so surrendering to the French held the risk of deadly retributions. Early that morning they had sent Wernher's brother, Magnus von Braun, down the road on the Austrian side of the border to find American soldiers and begin the process of surrender. To appear as nonthreatening as possible to any soldiers he encountered, Magnus had decided to ride a bicycle. In short order he had run into a small antitank platoon of the Forty-fourth Infantry Division.[4] Magnus had returned to Haus Ingeburg at 2:00 p.m. with news of his encounter with the Americans, telling his brother and the others that the American soldiers did not believe his story about a group of German rocket scientists holed up at a luxurious mountain retreat. He informed them that the soldiers wanted to meet some of the scientists in person, especially Wernher von Braun. They quickly developed a plan to send several of their group down the mountain to reassure the soldiers that Magnus had spoken the truth.

There would be seven of them in the advance surrender party. Wernher, Magnus, and Dornberger would be joined by Herbert Axster (Dornberger's chief of staff), Hans Lindenberg (a V-2 engine specialist), Bernhard Tessman (Wernher's special assistant), and Huzel, who had just returned from his clandestine mission in the Harz Mountains. Assuming the Americans would keep them for an extended period of time, the men packed some clothes into three cars, then followed Magnus's directions down the mountain.

At the bottom, the seven men encountered two jeeps full of well-armed GIs, who then escorted them to the Counter Intelligence Corps (CIC) headquarters in the town of Reutte. There they met First Lieutenant Charles Stewart who, owing to the lack of electrical power, was working at his desk by the light of a single candle.[5] Von Braun would later boast that he was not concerned that they would be mistreated. In an interview five years later he said, "I didn't expect to be treated as [a war criminal]. No, it all made sense. The V-2 was something we had and you didn't have. Naturally, you wanted to know all about it."

Once again, Wernher von Braun seemed to have a knack for predicting the future. He and his party were served meals far superior to what the GI grunts were eating, and in fact were treated almost like celebrities, shaking hands and posing for numerous photographs with the Americans. After his surrender, von Braun acted with so much bravado and gusto that one soldier quipped, "If we hadn't caught the biggest scientist in the Third Reich, we certainly caught the biggest liar!"[6]

In the end, approximately five hundred German rocket engineers would be questioned and interrogated by one or more of a half-dozen American intelligence agencies. Some of the scientists were deemed qualified to be allowed to enter the United States under the new top-secret technology transfer program, dubbed Project Overcast (later renamed Project Paperclip), while others were judged too steeped in Nazi dogma to be allowed to emigrate.

The officer assigned to question von Braun was a young by-the-book West Point graduate named Lieutenant Walter Jessel. For three days the lieutenant and the German rocketeer sat in a small room discussing von Braun's work in developing rocket technology. Assessing Wernher's psychological makeup was, of course, equally important. In his report, Jessel would later write a comment that would become prophetic: of all the German engineers, Dr. von Braun is "one of the most likely to be adaptable to American methods and ways of life."

At the end of the third day, Lieutenant Jessel had one last question for Wernher: "Dr. von Braun. If you were to emigrate to the United States and become a naturalized citizen, what would you do with the rest of your life?"

"I would build a rocket and fly to the Moon."

5

THE WILD, WILD WEST

We started wearing sombreros and cowboy boots. We started shouting yee-haw!

—Dieter Huzel

Bavarian waffles with blueberries. Apple raisin crepes with bacon. Cream puffs and brunch cake. Cheese and sausage strudel. These delicacies were on his mind as Wernher stared at his powdered eggs and stale toast. At Peenemünde they had been coddled—their every want and desire appropriated and delivered by Colonel Dornberger and the German military. The best food, the most comfortable furnishings, the softest bedsheets—nothing was too good for Wernher von Braun and his high-tech team. They had been generously paid and granted a research budget limited only by whatever they asked for.

Now, at Fort Bliss, the German engineers received, after deductions, a total of $4.80 per day, and were highly restricted on when, where, and how they could spend it.[1] They lived together in a barracks without air conditioning or heating. Rodents, scorpions, tarantulas, and rattlesnakes had to be routinely chased out of the building. There was no privacy, travel was prohibited, and their families had not been given permission to join them in the United States.

Worst of all, the beer tasted like mouthwash.

At his mess hall table sat Dieter Huzel, Arthur Rudolph, Eberhard Rees, and Maxe Neubert. Wernher knew what they were thinking—what all his men had been thinking since their arrival—did we make the right decision in leaving our homeland and coming to America?

The previous evening, they had needed to run an electrical power cord from one end of their barracks building to the other. While crawling under a bunk with the wire, one of Wernher's men had nearly been stung by a scorpion. Almost none of the buildings were air conditioned, power failures were common, and the Texas heat was unbearable.

Uncomfortable with the disconcerted stares from his colleagues, Wernher turned to look out the window to his right. From his chair in the uninsulated wooden dining hall, he could see the rocky, treeless Franklin Mountains to the north and the barren desert that surrounded them. For a man raised on pristine meadows and alpine forests, it was a stark transformation.

Wernher sighed, turned back to his powdered eggs, and bravely forced them into his mouth.

Von Braun's desire to be expatriated to the United States, though logical in retrospect, was anything but a done deal in 1945. The Americans wanted him, but so did many others, especially the British. Great Britain had suffered a great number of hardships and casualties during the war, and the U.S. government was sympathetic to British requests for postwar assistance. Winston Churchill even coined a phrase to describe this postwar battle for human science assets—the "Wizard War."[2] One of these "Wizard War" requests was to have von Braun assigned to Great Britain's newly created rocket research program. In a move that could have altered history in ways we can only imagine, von Braun was in fact sent to England soon after surrendering to the Americans in a British attempt to woo him. He and several other German engineers were temporarily interned at a prisoner-of-war camp in Wimbledon, not far from where the tennis championships are held.[3] While in England he was lobbied by Sir Alwyn Crow to stay, but von Braun's conviction that only the United States held the promise of bringing his Moon rocket dreams to pass caused that effort to fail. Crow then asked him to draw up a list German engineers who might prefer England over the States, which he did.

Before long the word came from America: Project Paperclip—the secret operation to bring a small number of Germans to the United States—had decided Wernher von Braun and a few of his associates were too valuable to be handed off or even lent out to anyone else. Permission to emigrate to the United States was granted. The travel arrangements required von Braun, ironically, to return to Germany for a short stay, then to continue on to Paris where, after a polite interrogation at a luxury resort, Wernher signed his first contract with the U.S. War Department. The contract granted him six months' work in the States, with a six-month option should the U.S. Army decide to extend this.[4] From

Paris there commenced a twenty-hour flight in a very loud, four-engine C-54 airplane that eventually landed in New Castle, Delaware. After a short trip to Boston in a smaller plane, von Braun and his military escort traveled by van and ferry to Fort Strong in Boston Harbor for more legal processing. Even after that processing, however, von Braun would still be considered an illegal immigrant under a strict interpretation of the law. In postwar America, the well-respected U.S. Army could get away with almost anything.

It was at Fort Strong that Wernher was given his final marching orders: he and a large contingent of his engineering team would be sent to Fort Bliss near El Paso, Texas, to engage in scientific research, the nature of which would be revealed later.

Von Braun was giddy—excited at the prospect of seeing the fabled "Wild West." Like most Germans, his familiarity with the American West extended from two sources: a few German-translated Hollywood movies, and the novels of German writer Karl May. May's American West stories revolved around a brave and heroic cowboy named Old Shatterhand and an Apache chief named Winnetou. Though a brilliant man, von Braun was utterly ignorant of how the nineteenth-century Old West had been mythologized by filmmakers and writers like May. Karl May, it turned out, had never set foot in America, having spent most of his life in Germany.[5]

Wernher von Braun was in for a surprise.

Hundreds of boxcars would be needed to transport the vast quantity of V-2 parts from American ports to El Paso. After arriving at Fort Bliss, von Braun discovered that those boxcars were strung out like a beaded necklace across the country to the harbors of New Orleans, New York, and Boston. Many more shipments were still on ships and barges on the Atlantic, and a large quantity of parts remained in Germany. The logistics of transporting so much cargo were immense, and many shipments became sidelined on spur tracks, forgotten or ignored.

Like the cargo they had designed and built, the same problem befell Wernher's fellow engineers as they became spread out across half the globe. Some remained scattered throughout Germany and several European countries as they underwent military processing and further interrogation. Others awaited permission to travel, while some were simply stuck in transit due to postwar transportation dilemmas like fuel shortages. Due to his unarguable importance, Wernher's travel arrangements were given top priority and he was in the first group of German engineers to be sent to Fort Bliss. Immediately upon arrival he became bedridden with hepatitis and ended up in the uncomfortable position of sharing a hospital barracks with GIs who had been wounded in the war by weapons of his design.[6]

Over time, von Braun would be joined at Fort Bliss by more than one hundred of his former associates.

Von Braun knew his V-2 hardware was being shipped to Fort Bliss, and he naturally assumed he would be put right to work assembling and testing V-2 rockets. The Americans did not have his level of rocket technology, and it was only logical to expect that they would want him to be involved in helping them learn it. Instead, he was surprised to find himself assigned to work on a top-secret, next-generation propulsion system: the ramjet engine. Despite knowing nothing about the technology, von Braun was one of several scientists put in charge of its development. Since ramjets would require burning atmospheric oxygen as one of their propellants, this assignment meant he would be working on a technology doomed to never go into orbit, let alone fly to the Moon. For von Braun, dreamer of lunar landings, it was yet another setback—another example of government-funded myopia. Was no one else aware of his true destiny? After arriving in Texas, it took only a few weeks for him to grow concerned that he was already becoming superfluous to the Americans.

Eventually, though, reality hit the military brass of Fort Bliss like a ballistic missile. As V-2 parts started accumulating everywhere on base (they filled all the available warehouses and overflowed into numerous outdoor storage yards), it became obvious that von Braun's expertise would be required elsewhere. The Army needed people who could maintain, assemble, fuel, test, and launch the V-2 rocket, and for that task there was only one man. But there was another concern. Though the war in Europe was over, the United States was still at war with Japan—the atomic bomb had not yet been dropped on Hiroshima. This factored into the U.S. Army's decision to allow von Braun and the Germans to focus their Fort Bliss work on the V-2 ballistic missile.[7] It was possible they might need such a weapon themselves.

Putting the Germans to work on the V-2 meant frequent visits to the launch range across the New Mexico border. Though life at Fort Bliss could be boring, uncomfortable, spartan, and downright miserable, Wernher and his men were in for a shock. If they thought Fort Bliss was bleak, it was paradise compared to what was yet to come.

The White Sands Proving Grounds lived up to its name—a seemingly endless swath of gypsum-based white sand that covered everything and got into anything. After the German scientists began assembling V-2s in preparation for test launches, they soon discovered a great deal of extra time had to be invested in cleaning sand out of the rocket hardware—the engines, propellant tanks, fuel lines. The sand also bedeviled pretty much every orifice in their bodies. White Sands was an ethereal wasteland to end all wastelands, and working there only succeeded in increasing the level of unhappiness of America's new rocket scientists.

But it wasn't just the sand that made the experience hellish—the facilities on base were so meager and unprofessional, it drove the engineers mad. In one famous moment they would talk about for many years, von Braun had instructed the Americans that they would need a safe and protected station not far from the launch area to observe the rocket and perform an emergency abort if they observed something they thought was hazardous (overflying airplanes, unexpected visitors, errant flight path of the rocket, etc.). On a May day in 1946, they were scheduled to launch the first V-2 from White Sands. Von Braun was shown what the Americans had prepared for their abort station: a skinny wire strewn along the ground from the launch pad, connected to a small switch—there was no building or structure of any kind. The only protection from explosions or falling rockets was a small sand dune. The Americans told von Braun that this was his "Emergency Cutoff Station."[8] According to Dieter Huzel, the missile launching infrastructure of White Sands was, at times, "no better than our unfunded operation as amateur teenagers back in Germany. The Americans wanted results but had no budget for anything we were being asked to accomplish."[9]

Though seemingly amateurish, the abort switch ended up being needed that day, as the first V-2 launch went horribly awry. Wolfgang Steurer, one of the German engineers working beside von Braun, described what appeared to be an utterly wild flight. According to Steurer, the rocket, "tumbled happily in all directions."[10]

A month later the Germans scored their first major success with the V-2, launching their rocket to an altitude of sixty-seven miles—an important milestone, as the beginning of nonatmospheric space is considered to start at fifty miles. The rocket even carried a significant payload—an upper-atmospheric instrument package, first suggested by Arthur C. Clarke more than a year before. Some people began referring to that April 1946 launch as the "dawn of the Space Age," but a better candidate for that title would occur a few years later.

The United States had not been entirely sedentary when it came to high-performance liquid fuel rocket technology. America's Robert Goddard was, after all, the pioneer in the field, having launched the world's first liquid fuel rocket near Auburn, Massachusetts, on March 16, 1926. By the time 1945 rolled around, several companies, including Douglas Aircraft and the Guggenheim Aeronautical Laboratory, had built rockets, both solid and liquid, with significant altitude successes, one of which was the WAC Corporal. Von Braun, frustrated with the inability of the U.S. Army to assign him to any significant new project, decided it might be interesting to see what would happen if a WAC Corporal was mounted as a second stage atop a V-2. The Chinese had invented the idea of rocket staging all the way back in the fourteenth century, but the concept had yet to realize its true outer space potential. On February 24, 1949, a V-2 rocket with a WAC Corporal second stage lifted off from the White Sands Proving Grounds. The rocket scored several firsts: highest speed ever attained by a man-made object—5,150 mph (more than five times the speed of sound), and the highest altitude ever recorded—250 miles. It was an astounding achievement, completely unnoticed by the media at the time.

That humans could build and launch a device that could defy Earth's pull and rise to such an incredible height should have changed everything. How could there be any more doubt? Wernher von Braun was right—all you needed to get into orbit, or fly to the Moon, was a rocket big enough to get you there.

Still, no one paid attention.

Wernher finished his powdered eggs and stale toast, then looked quietly around the table at his close friends and associates. Another ride over dusty, rutted roads awaited them that morning—a launch was scheduled at White Sands—and he could read the lack of enthusiasm on their faces. Outside in the Texas sun their bus awaited, its driver impatiently honking the horn.

"Lass uns gehen," said Wernher.

Everyone stood up, and the men followed him to the door.

6

MAN IN SPACE

There is nothing like a dream to create the future.

—Victor Hugo

The F9F Panther fighter jet, built by Grumman Aircraft, was no match for the far superior and much faster Soviet-built MiG-15, and its pilot kept a close eye on his airspace and a closer ear to the radio chatter, lest any MiGs be detected in his vicinity. The pilot had reason to be more concerned than normal. He had just dropped his last bomb, hitting its target like a bull's-eye and destroying a key enemy bridge. But his plane had a major problem; just after releasing the bomb, the jet fighter sheared its starboard wing on an antiaircraft suspension cable, reducing the length of the wing by more than six feet, and greatly reducing the jet's airworthiness. The plane's gas tanks were still more than half full, it was carrying hundreds of pounds of munitions, and at 350 knots it was only five hundred feet off the ground—the perfect setup for a deadly, fiery crash.

The pilot had taken off many miles away, part of a fleet of planes deployed from the U.S. aircraft carrier *Essex* off the coast of Korea. The fighter plane was now far too damaged to risk a return over many miles of ocean back to its home ship, so the pilot began considering his options. He still maintained enough aileron control that he could slowly gain altitude, and that was his first maneuver. Beyond that, his choices were few and dangerous. He needed to get up to about 14,000 feet in elevation where the air was thin enough to increase the odds of surviving an ejection, and he needed to leave communist Korean territory, lest he parachute into enemy hands, get captured, and be executed. Still, ejection was not necessarily the best option. The pilot had flown with Chuck Yeager at

Edwards Air Force Base, and the famed aviator's words now played out in his mind. According to Yeager, ejecting from a speeding jet was akin to "committing suicide to avoid getting killed."

The pilot had radioed his dilemma to his fellow jet fighters, and one of them, John Carpenter, now maneuvered his jet alongside. "We'll make it," was Carpenter's simple advice. Then, in preparation for the ejection, he added, "Make sure your shoulder straps and seat belts are tight." Carpenter stayed with the pilot until they were out of enemy territory, then flew off, knowing that once the ejection seat fired, any plane in close proximity could be hit by flying debris. Now alone, the pilot waited a few more seconds, then actuated the ejection seat. At high speed and ultrahigh g-forces, the pilot's body was fired into the air. The drogue chute deployed, followed by the main, and the pilot soon found himself descending gently downward. To his left, the vastness of the Pacific Ocean spread out to the horizon. To his right lay the lush green landscape of Korea. The slow descent took several minutes, and the pilot half expected to be shot out of the air any moment by ground fire or a passing MiG. Instead he landed gracefully in a rice paddy, miraculously uninjured.

Doffing his parachute and cracked helmet, the pilot heard a sound—a sound he was quite familiar with. It was the purr of an approaching U.S. military Jeep, sent from a nearby Marine base. Standing in a foot of water, he looked up to see a familiar face at the wheel—his former roommate from flight school, Goodell Warren.

As the Jeep pulled up close, the youthful twenty-one-year-old pilot flashed a toothy grin and called out to his friend, "Goodie—you never looked so good!"

Goodell parked the jeep, stepped out, and faced the pilot.

"Neil Armstrong—what the hell are you doing in my rice paddy?"[1]

For five years Wernher von Braun and his German expatriates had worked for the U.S. Army as "special employees," a euphemism for illegal immigrants.[2] One day the U.S. government decided the time had come to normalize their status and instructed the Germans to do something very odd: walk across a bridge that extended from El Paso, Texas, over the Rio Grande River, and into Mexico, then turn around and return to El Paso and U.S. soil. This technical maneuver of exit and re-entry was deemed necessary in order to begin the process of permanent legal residency. After five years in the States it was fairly clear to everyone—the Germans were here to stay.

Soon after that bureaucratic detail was completed, the rocketeers received word that the space they occupied at Fort Bliss would have to be vacated due to military needs involving the ongoing Korean conflict. The base would no longer have space for experimental rocket testing. At about the same time, the Pentagon was concerned about intelligence reports stating that the Soviet Union was building nuclear-capable intercontinental ballistic missiles (ICBMs). Suddenly the Germans were in demand. Two weeks later, when space opened up at a larger facility in Huntsville, Alabama, the decision was made to move what was now being called the "Redstone Arsenal" and most of the German scientists to that location. Curious as to what lay ahead, Wernher took a trip to Huntsville to tour the facility. Upon returning to El Paso he was brimming with enthusiasm, like the Wernher von Braun of old. "Oh it looks like home!" he exclaimed. "So green, green, everything is so green, with mountains all around."[3] This was a relief to all the men who had grown up amid the verdant beauty of Germany. As Hannes Luehrsen, von Braun's architect-planner, would later remark, "At [Fort] Bliss we had to drive two hundred miles to see five trees together."[4]

With the move, Wernher von Braun received a new title: Technical Director, Army Ordnance Guided Missile Development Group, Redstone Arsenal, Huntsville, Alabama.[5] The men, assorted clerks and technicians, and many trainloads of equipment soon settled into their new digs. They were energized not just by the new location, but by a new mission: the Pentagon had finally caught the vision of rockets as a modern-day necessity—tanks and battleships were not going away, but this new invention had to be exploited to ensure superiority, or at least parity, to what was going on in Russia.

After settling in Huntsville, the news of the existence of the German rocketeers could no longer be kept secret. At Fort Bliss the men had been required to shop at the base commissary and discouraged from going into town in El Paso, lest their identity and presence cause a public-relations problem for the army. Even so, von Braun encouraged his men to fraternize with the American people in town whenever possible. A few of his engineers ended up meeting young American women and getting married. After the move to Huntsville, the army decided to loosen its leash and allowed the Germans much more freedom in hobnobbing with the local gentry. Von Braun continued to encourage a policy of outreach, realizing that being friendly with the locals was a key to their acceptance, and therefore their success. Though there were a few cases of shopkeepers putting out signs with slogans like "NO KRAUTS ALLOWED," the interaction between the Germans and the people of Huntsville eventually evolved from strained to accepting to friendly.

As he had at Fort Bliss, Wernher began to receive invitations to speak before various community groups. One evening he was at a dinner hosted by a fundamentalist religious group. One of the pastors excoriated von Braun in front of everyone, declaring, "This drought we've had in Alabama these past two years has ruined our crops! When are you going to stop punching holes in the clouds with those rockets and drying up the rain?" Von Braun's response showed how well he had his feet planted in two worlds simultaneously—the world of science, and the world of religion. "I know you are familiar, sir, with the Bible and with the story of Jacob's ladder," replied Wernher. "The angels are ascending and descending the ladder. So are we. If the good Lord does not want us to go up and down His creation, all He has to do is tip over the ladder." The audience gave von Braun a deafening applause, assuring him a positive reputation.[6]

Before long the presence of the German engineers became widely known, and requests for media interviews became common. Wernher was a busy man, and in order to save time he decided to hold a press conference. He expected it to be a small affair, but to his surprise journalists from all over the country showed up. He fielded a wide range of questions, but one in particular was the most memorable. Near the end of the press conference a reporter stood up and asked, "What would it take to send a man to the Moon, and return him safely to Earth?" The question made the roomful of quiet, respectable journalists laugh out loud. They quickly grew silent, though, as they noticed von Braun's serious expression. For Wernher there was no greater question than the one involving manned flight to the Moon, and it was imperative that he answer it correctly. He cleared his throat, quietly gathered his thoughts, then spoke his answer. "The only thing it would take to send a man to the Moon and return him safely to Earth, is the will to do it."

As the journalists scribbled his response in their notebooks, the German scientist continued. He assured the reporters that everything was now in place to send a man into space. It would be risky, but the risks were worth the payoffs. "America is ready," he said. "We have the best scientists, and the most advanced rockets." Secretly, however, he questioned the truth of that statement. Intelligence was beginning to leak out of the Soviet Union, and the reports all were in agreement: the Soviets were working on something big.

It was midnight on a Sunday evening. In Huntsville, Wernher von Braun spent his days performing his job for the U.S. Army: designing rockets for the present. By night he engaged in his favorite hobby: designing rockets for the future. If anyone had bothered to come by the Redstone Arsenal mess hall and ask Wernher what time of day it was, he probably would not know the answer—he had a clerk and a secretary to keep track of such mundane details. People with drive, ambition, and focus often care little what date is circled on the calendar, or what numbers a clock's hands point to. All that really matters for such people is what is in front of them, and what was in front of von Braun on that late winter evening in 1952 was Mars. At a time when almost everyone on Earth believed sending a man-made satellite into space was impossible, that human spaceflight was outlandish, that landing humans on the Moon was silly science fiction, Wernher was already performing the calculations necessary to send a manned spaceship to Mars. His incredible vision, his uncanny ability to think far into the future, would pay off in ways he never expected.

Wernher's Mars work was more than just idle theory, however. His press conference had a much wider impact than he expected. Requests for articles and speaking engagements were pouring in. Of all the offers he received, the one that had intrigued him most was from *Collier's* magazine. *Collier's* wanted to do a series of articles over two or three years focused on human spaceflight, and they wanted Wernher von Braun to write the first one (if not more). In addition, a professional artist would be hired to paint realistic-looking pictures of what the space stations and spaceships of the future would look like, and those pictures would be published as part of the articles.

And best of all, it included a paycheck.

It was far too good an offer to pass up, so on that late evening Wernher was working on what would become *Collier's* first manned space article. At Wernher's urging, the magazine also secured the talents of Willy Ley, a writer and close friend of von Braun who had snuck out of Germany prior to the advent of World War II. He had done so by using a German-approved journalism visa that allowed him to travel for a few weeks to the United States, and then return. Only Ley never returned.

Wernher's first article for the magazine was entitled "Crossing the Last Frontier" and would be published in the March 1952 issue. The issue included fanciful color drawings of massive spaceships, some with wings, foreshadowing Space Shuttle–like reusability. The tagline at the top of the magazine's cover prophetically read, "Man Will Conquer Space Soon." Over the next three years, the articles and illustrations, enhanced by von Braun's imagination, would galvanize the attention of Americans everywhere. Von Braun explained in simple, easily understood terms how spaceships, orbital space stations, and flights to the Moon and the planets could be accomplished quickly. The technology was already on the shelf—all that was needed was public support and funding.

As soon as the first set of articles was published, Wernher's imagination caught the attention of Mr. Imagination himself: Walt Disney. The animation inventor was about to open the world's first theme park. One of the areas of his park was called Tomorrowland—a showcase of all things yet future. Disney needed help, however, to solidify the details of that future, and he saw in von

Braun a kindred spirit—a man who shared his idealistic view that what the human mind could imagine, human hands could build. Disney contacted von Braun and asked for his assistance in a new venture. Disney needed someone with a "Tomorrowland mentality" to appear on his new TV show, *Man in Space*, and explain in front of a TV camera what von Braun had described in the magazine. Von Braun—ever the happy showman—readily agreed. *Man in Space* was a success, so Disney followed it up with two sequels: *Man and the Moon* and *Mars and Beyond*.[7] By the time the third show had aired, Wernher von Braun had become a celebrity, and Disneyland was a smashing success.

Americans were quickly getting fired up about the Space Age—an era of human endeavor that was being talked about before it even existed. Despite his over-the-top enthusiasm and penchant for hyperbole, von Braun was right about one thing: most of the technology needed for manned space flight already existed. Now, with the American people quickly getting on board, and the manufacturing infrastructure in place, all that was needed was for Washington politicians and the U.S. Treasury to get in line.

7

CHIEF DESIGNER

There is no such thing as an unsolvable problem.

—Sergei Korolev

The competition between the United States and the Union of Soviet Socialist Republics that would soon be referred to as the Space Race got its start, albeit unwittingly, in the humble Iowa City living room of a physicist named James Alfred Van Allen. It was March 1950, and the Cold War was getting hotter, largely due to the unstable, paranoid rantings of Joseph Stalin. Several major U.S. scientists including S. Fred Singer, Lloyd Berkner, and Sydney Chapman gathered at Van Allen's home to discuss the possibility of cooling the Cold War by organizing a friendly worldwide endeavor in which many nations would work and collaborate together in the common interest of scientific discovery. It would last twelve months and would be called the International Geophysical Year. A proposal was made to the International Council of Scientific Unions to sponsor the event, and the council agreed. After much planning and discussion, it was decided to schedule the IGY to coincide with an expected high degree of solar activity several years away and expand it from twelve months to eighteen, commencing July 1, 1957, and running through December 31, 1958.

A few years after that living room discussion, on July 29, 1955, James Hagerty, President Eisenhower's press secretary, announced that the United States, as one of its IGY projects, would launch the world's first artificial satellite. A mere four days later, at the Sixth Congress of the International Astronautical Federation, Leonid Sedov—a scientist from the USSR—announced that the Soviet Union would do the same. No one spoke of any sort of competition;

the scientists had no desire to sully their reputations by promoting something so juvenile as a race. But whether they realized it or not, both countries had just thrown down the gauntlet—the Space Race had begun.

No one was more frustrated about this unofficial race than Wernher von Braun—one of the few who understood the race had begun long before the Van Allen living room summit. Von Braun was frustrated because he knew that, if put in charge, he would win the race. Yet within the U.S. space program, he was still far down the totem pole of policy makers. On the other hand, no one was more pleased with the challenge than rocket engineer Sergei Korolev. He was ecstatic, given that he *was* in charge of the Soviet program, and already had on the drawing board heavy lifting rockets capable of launching orbital payloads. Immediately after Leonid Sedov's announcement, Sergei made it his mission to make certain the USSR was first in the satellite race.

Born in the Russian town of Zhytomyr, in what is today Ukraine, Sergei showed an interest and talent in mathematics at a very early age. He attended the Kiev Polytechnic Institute for a short while before transferring to, and graduating from, Bauman Moscow State Technical University with a degree in aviation design. Korolev looked to have a bright future, but under Stalin's rule nothing was certain. After being falsely accused of treason by a close associate, Korolev, along with hundreds of other scientists and engineers, was convicted and sent to a slave labor camp. Sentenced to eight years, he was released on June 27, 1944, having served six years of his sentence. While in the gulags and labor camps, Korolev had been assigned design work on planes, bombers, and rocket engines—experience that would bring to pass his destiny.

Like Wernher von Braun, Sergei Korolev began to envision the possibility of using large rockets to fly humans into space, into orbit, onto the Moon. Due to his aviation engineering degree, along with his wartime experience with rocket engine design, Korolev was one of the Russians assigned to travel to Germany at the end of World War II to collect as much German technology—and as many German scientists—as possible. He and the other Soviet "recruiters" managed to capture two thousand German scientists and technicians and collected thousands of tons of aviation and rocket hardware, most of which was transported back to Moscow and other Soviet destinations.

Korolev's leadership skills were evident during his trips to and from Germany, so Joseph Stalin decided to put him in charge of liquid fuel rocket design. Korolev wasted no time in his new profession. He assembled an able team of engineers, technicians, and metal-benders and went right to work. His first self-imposed assignment was to take the V-2 blueprints obtained from a captured German engineer and reconstruct and flight test a number of V-2 rockets. After successfully completing that, and learning a great deal from the experience, Korolev ordered the design and creation of a succession of larger and ever more powerful rockets, culminating in the designing of a rocket he named the R-7—up to that point the largest and most powerful rocket ever conceived. It would be the rocket that would eventually lift Sputnik into orbit.

There was no doubt in Stalin's mind, nor that of his successor, Nikita Khrushchev, that Sergei Korolev had quickly become one of the Soviet Union's most important assets. At one point the communist leadership began to fear the possibility of an assassination attempt by the United States, so Stalin ordered Korolev's identity to be kept a state secret—no one was allowed to speak his name, or even address him in person by his name. Everyone who knew of or worked with Sergei Korolev was instructed to refer to him by the title "Chief Designer." Saying or writing his name could get one sent to the gulag.[1]

The papers and blueprints on his desk fluttered as the door of Sergei's wooden-shack office opened and the Siberian winter wind, along with a few snowflakes, blew in.

"Chief Designer."

Sergei did not look up. "Close the door."

"Yes, Chief Designer."

Sergei finished the slide rule calculation he was making, jotted the answer on paper with his pencil, then glanced up from his work. There was a young man—a boy really—standing in front of him holding an armload of firewood.

"Chief Designer—I brought you more wood."

Sergei pointed to the brick hearth to his right and the boy gently placed the wood there.

"Shall I put some on?"

Sergei nodded, and the boy placed three cut pine branches onto the bed of red coals.

"Chief Designer. May I get you anything else?"

Sergei shook his head.

"Some coffee? Bread? I hear we have a little butter today."

"No, thank you. Now you may go."

"Yes, Chief Designer."

The boy left, closing the door firmly behind him.

Sergei set his pencil down and rubbed his eyes. He had no idea of the time, but was certain it must be after one in the morning. Perhaps after two. He and his engineers had a heat transfer problem with their new liquid fuel rocket engine and the solution was avoiding them. Sergei looked to his right and noticed the three American magazines dropped off that morning by officers from the KGB. Beginning with the March 1952 issue, they were the first three issues of *Collier's* dealing with Wernher von Braun's vision of manned spaceflight. Along with the magazines, the KGB had also provided English translations of all the space-related articles. He had read them immediately, of course—to Korolev, information about what was going on inside Wernher von Braun's head was more valuable than gold. Sergei had marveled at what he read. Von Braun's theories and proposals were brave, bold, astounding. While Sergei just recently had begun considering the possibility of sending men to the Moon, von Braun was already planning missions to Mars! It was beyond belief, and could mean only one thing: with Wernher von Braun at the helm, the Americans must be well along in planning the world's first artificial satellite launch.

The problem was, Sergei Korolev had already promised that honor to himself and the USSR.

The forty-watt lightbulb hanging above him dimmed, and outside Sergei could hear the generator slow and sputter. It continued like that for several seconds, then resumed its normal purr, allowing the bulb to brighten. He turned to glance at the far wall with its large blueprint pinned up, prominently displaying the overall design of the R-7 rocket.[2] The R-7 design consisted of a large core rocket with four large disposable boosters strapped around it. Fully assembled, it would be thirty-four feet wide at its base—one foot wider than the yet-to-be-designed Saturn V—and stand nearly one hundred feet tall. To save money, it would cluster a total of twenty engines, much smaller, and therefore less expensive, than those that would soon be developed by the Americans. The engines burned kerosene and liquid oxygen—a combination that would one day be replicated by the Americans on the Apollo program. Fully fueled, the R-7 would weigh 280 tons and be capable of lifting into orbit a payload of 5,000 kilograms. It was the kind of rocket Wernher von Braun would build—if the Americans would let him.

Sergei turned his attention to the shack's small window, and the thermometer resting just outside the glass. Its mercury indicated a temperature of −13° Celsius.

Thus far, the only official word from the Americans regarding space goals was their intention to launch a satellite during the coming International Geophysical Year, still more than four years away. Sergei's handlers within the Politburo had wanted to keep their own satellite goals secret, but once the Americans had announced, international pride was at stake and so the Soviet leaders felt they had no choice but to respond in kind.

Sergei paged through *Collier's* magazine, marveling again at the drawings that illustrated the plans and dreams of Wernher von Braun—the pictures and descriptions were inspiring. Sergei Korolev, however, did not understand the difference between American literary entertainment and actual project planning. He assumed that at least part of what he read in *Collier's* was also on some drawing board somewhere, marching toward fruition. Ironically, his misunderstanding would strengthen his resolve to not only continue working toward his own goals, but to redouble his efforts and achieve them before his perceived adversary. The Soviet Union's Chief Designer set the magazine down, picked up his slide rule, and began his next calculation.

8

COMPETITION

It's a difference of opinion that makes horse races.

—Mark Twain

The Soviets were not the only rocket builders forcing von Braun to look over his shoulder. Within his newly adopted country there were two other rockets competing with his Redstone for orbital launch attention. Unbeknownst to von Braun, his drumbeating over the importance of furthering rocket research had not landed on deaf ears, as he had assumed. After submitting his postwar report to his American handlers about the importance of this new technology, both the U.S. Navy and the U.S. Air Force soon launched their own secret programs.

Beginning in 1946, the Naval Research Laboratory began design and construction of a sounding rocket originally called the Neptune, but later renamed Viking. Though half the power and weight of the V-2, almost every major aspect of the Viking's design was copied from von Braun's rocket. Two of its designers, Milton Rosen and Ernst Krause, initially visited von Braun at Fort Bliss to pump him for design ideas and information. The German scientist freely answered their questions, providing whatever technological details they required, but the Naval Research Laboratory refused to reciprocate, deciding to leave von Braun out of the loop as their project progressed. But once the Viking rocket was constructed and began undergoing flight testing at White Sands, concurrently with the V-2 launches, it became impossible to keep the details secret from von Braun and his German engineers.

The slight of being excluded from the Viking project must have bothered the Germans, given that they were the authors of so much of its technology. But the indignities would only get worse. By the time the 1950s arrived, the navy was in the serious planning stages of taking the Viking rocket and setting two stages atop it to form a satellite launch vehicle. The satellite-capable incarnation of their rocket was dubbed Vanguard, and von Braun immediately recognized it for what it was: a direct competitor of the U.S. Army's Redstone. Again, he and his engineers were left out.

For Wernher, the Viking and Vanguard rockets were yet more proof of how the Americans did not take him seriously. Unfortunately, this lack of confidence in his abilities, coupled with the feared political fallout from his involvement, were hurdles partially of his own making. Wernher von Braun was a dreamer of big dreams—too big, as it turned out. In the October issues of *Collier's*, von Braun described a near-space future so audacious that there would not be enough money to fund it even if every country in the world contributed. He declared the first human landing on the Moon would include three spaceships with a total crew of fifty men. The ships would be built in Earth orbit over a period of six months at a space station (which, of course, would also have to be built). Each ship would be 160 feet long and 110 feet wide.[1] The capsules housing the astronauts would be five stories high. Once on the Moon, one of the three ships would be turned into a permanent Moon base. After exploring the Moon in several roving craft, the fifty crewmen would return to Earth in the other two spaceships.[2] It was a plan so grandiose as to tarnish, for a time, his reputation within the budding space industry, and fueled the idea that von Braun should not be allowed to have any part in planning American space policy, lest he bankrupt the U.S. government.

But even on the ground, in the world of more pragmatic and practical rocketry, von Braun had good reason to be concerned about his future influence and participation. The Viking/Vanguard had several design innovations that, over time, proved to be major improvements over the German's rockets. In order to save weight, and thereby increase performance, the propellant tanks were integrated into the fuselage—not separate tanks, like the V-2, which required a weighty superstructure to hold them. In addition, the navy's rocket utilized gimballed rocket motors—the rocket was steered by swiveling the rocket engines. The V-2 had used deflection vanes that extended into the rocket's exhaust—a design that was less efficient.[3]

Not to be outdone, the air force jumped into the fray with its own design. In a move that seemed to imply distrust of the army's Redstone rocket, the air force began working on its own iteration of an ICBM. Built by Convair, it was named after Convair's parent company: Atlas. The Atlas rocket shared a major design detail with the R-7. Like Korolev's rocket, the Atlas had a core rocket that would fly from takeoff to final destination. Instead of tandem stages, it utilized several strap-on booster rockets that were discarded in the upper atmosphere.[4] The air force decision to jump into the rocket business proved critical for human spaceflight, for although Wernher von Braun's preferred tandem staging design would eventually win over converts for the future Moon missions, it was the Atlas that would be tapped to send the first American, John Glenn, into orbit. In fact, not only would the Atlas be used to launch the last four Mercury 7 astronauts, it was also destined to become the workhorse of the future U.S. satellite business.

Sitting in his office in Huntsville, Wernher could feel his space travel dreams beginning to unravel—again. Apparently being the inventor of large-scale liquid fuel rocketry was not enough—the circus was moving on without its ringleader. Intelligence reports crossing von Braun's desk indicated the Soviets were clearly building the infrastructure necessary for massive rockets and were doing so at a frenzied pace. At home, he had to contend not only with his increasingly disgruntled German colleagues, but also competition from the well-budgeted navy and air force rocket programs. Wernher could not understand the army's odd inability to receive funding on a par with those two branches. Had he hitched his wagon to the wrong star? Though it was true that the air force Atlas program was at least two years behind the Redstone, it was also clear that in the postwar world, the air force budget requests were being given a higher priority over the army. The Atlas would soon be catching up to, and passing, the Redstone.

In desperation, von Braun considered how to circumvent the slow-as-molasses forces that were holding him back. He wondered if it might be possible to cobble together existing army rocket programs that he had access to and mold them into a viable satellite-launch vehicle. He had the booster: the Redstone main stage. But what could he put atop it in order to push a payload high enough (at least 150 miles) and fast enough (17,000 mph) to obtain orbit? The Army had in its arsenal a small solid fuel ballistic missile called the Loki. Only a few inches in diameter, it was meager by any standard. But in his urgency to do something—anything—von Braun began to formulate a desperate orbital launch plan. The idea was to create a four-stage rocket using the Redstone as the first stage. The second stage would be a cluster of nineteen Lokis, the third stage a cluster of seven Lokis, and the fourth stage would use three Lokis with a small satellite mounted above. It was a complicated mishmash filled with numerous possible failure scenarios, but it could be done fast, and on the cheap. On June 25, 1954, von Braun and his staff made a formal presentation to a group of navy and army brass to build and launch his proposed down-and-dirty satellite rocket, and the idea was in fact carried enthusiastically up the chain of command for a while. Eventually, however, more sensible heads—and interagency politics—prevailed, and the idea was abandoned.[5]

In Washington, ensconced in the Naval Research Laboratory, Vanguard Project Director Milton Rosen could also feel the hot breath of competition on his back. He had joined von Braun and Korolev as one of the few individuals who understood that a real race was underway, and like the other two men he was determined to win it. The Vanguard, he promised himself, would carry the world's first satellite into orbit. With backing from the American Rocket Society, General Electric, the National Science Foundation, and, of course, the U.S. Navy, Rosen felt he had everything in place necessary to compete in the space satellite race and win.

What he did not have was patience and real-world rocket experience, as a series of major Vanguard failures would soon reveal.

At the same time, unbeknownst to almost everyone in the United States, the Soviet Union was experiencing its own internecine rivalry between individuals, departments, politicians, and military branches. Despite Korolev's many proven successes, several other groups of rocket advocates had Khrushchev's ear and were siphoning off large quantities of rubles for their own projects.

Mitrofan Nedelin had been awarded the Hero of the Soviet Union title for his distinguished military career during World War II. After the war he quickly followed an interest in rocketry and was later influential in convincing Soviet authorities that the Nazis had had the right idea, just not enough time and money. Large rockets were the best way to carry warheads across large distances and the globe to Russia's enemies, he claimed, and before long Nedelin held sway over how some rockets were budgeted.

Though publicly Khrushchev would self-congratulate the USSR whenever Korolev's rockets scored a success, behind the scenes he was unhappy that Korolev had chosen a design for which the propellants were not storable for long periods. All of Korolev's rockets used liquid oxygen, which has a very short shelf life. The drawback of using nonstorables was that Korolev's rockets, no matter how grand, could not be used as efficient ICBMs. To be effective as an ICBM, a rocket had to stand loaded and ready for months and years on end, which meant it required storable propellants. Khrushchev, who thought he had been financing weapons, was beginning to feel he had been snookered by Korolev, which, to some degree, was true.

Enter Mikhail Yangel. A former understudy of Korolev, Yangel was a rocket engineer who had been promoted over Korolev in 1952 to the NII-88 research institute—an important post that would govern missile and space policy. Yangel subsequently became a close advisor to Khrushchev after he presented an ICBM design that could use storable fuels—hypergolic propellants[6] that included forms of hydrazine as a fuel and nitrogen tetroxide as an oxidizer. He used his influence with Khrushchev to guide how rocket money should be spent, with himself one of the key recipients. Soon Yangel was diverting large amounts of Russia's rocket budget away from Korolev, which drew Sergei's ire, and forced him to spend more time lobbying the Politburo. We will never know what mankind might have achieved in space had it been more focused on scientific discovery during this period than on weapons-related political rivalry.

Eventually fate would play a hand in awarding Korolev more control over rocket research dollars, and would do so in spectacular, and deadly, fashion.

Nedelin and Yangel's project lagged far behind the promised schedule. In the spring of 1960 Khrushchev chastised the men for not delivering the storable-propellant ICBMs they had promised. Work was going too slow, and he leveled a veiled threat if there wasn't progress, and soon. Khrushchev demanded a demonstration launch of what the two men were calling the R-16 rocket, and required it be launched by "the following autumn." Nedelin and Yangel knew their rocket would not be ready, but there was no arguing with the premier. They pushed ahead, made shortcuts, took a few calculated risks, and scheduled the maiden launch for October 23.

In the run-up to the launch, Nedelin and Yangel discovered what the Germans already had known for years—acidic oxidizers are not truly storable for long periods. They corrode the best materials, they leak through the best valves, they bring innumerable toxic headaches. The promises the men had made to Khrushchev were an illusion. As two hundred men worked around the clock at the Baikonur launch complex, a leak in the oxidizer side of the system appeared, and Nedelin ordered the men to work throughout the night to make repairs. To save time, he ordered the repairs be performed with the rocket on the launch pad, its propellants still fully loaded in their respective tanks. Proper procedure required the tanks to be offloaded before such maintenance, but Nedelin was focused on launching as fast as possible. Offloading the hazardous propellants, only to have to reload them again later was, in his mind, a waste of valuable time. In order to make sure the ground crew did not dawdle, he set up a stool thirty meters from the launch pad and sat there, watching everyone closely and making sure no one slacked off. Konstantin Gerchik, a major general in charge of the Baikonur Cosmodrome, urged Nedelin to retreat to a safer position. The situation was highly dangerous, but according to Gerchik's memoir, and Soviet rocket lore, Nedelin was without fear, telling Gerchik, "What is there to be afraid of? Am I not an officer?"[7]

In the morning, Nedelin signed off on the flightworthiness of the R-16 and ordered the crew to commence the launch countdown. "T-minus-zero," however, never arrived. Thirty minutes before the scheduled launch an incorrect command was sent to the upper stage, igniting its engines. That fireball ignited the propellants of the second stage mounted below it. Then, like dominoes, giant balls of explosive fire cascaded downward, engulfing and obliterating the rocket and everything within a one hundred meter radius. Yangel managed to escape the inferno—he was in the bunker having a cigarette at the time. But Marshal Mikhail Nedelin, chairman of the State Committee, was not so lucky. Sitting in his stool thirty meters from the rocket, his body was vaporized. All that was left of him was the collection of war medals that had been pinned to his uniform.[8]

It would be years before news of the accident reached the United States. The official death toll was put at 92, but eyewitness statements say the actual death toll should be closer to 150.

As for Korolev—he just kept building rockets. And with Nedelin now out of the way, Korolev's budget soon loosened up.

Joseph Kennedy Sr. had made his fortune in the film business. Prior to World War II his intention had been to leverage his influence and wealth to propel his eldest son, Joseph Kennedy Jr., into politics. The patriarch of the Kennedy clan, much like Wernher von Braun, had dreams and goals that were nothing less than audacious. One of those goals was to elevate his son to great political power—to help him become the first Catholic president of the United States. But fate intervened, and Joseph Jr., a pilot in the Pacific during the war, perished in a crash in 1944. The father grieved for a long time but managed a clear enough head to switch gears and favor son number two. In 1946, soon after John returned home from the Pacific, Joseph Sr. used his influence with the Hearst Newspapers to get John a journalism position—a job that would afford his son a high public profile.

That job, however, was short-lived. Once again using a level of influence that might be considered unethical today, he convinced James Curly, congressman of the 11th Congressional district, to step down in favor of becoming the mayor of Boston. In the primary election that followed, eleven Democratic candidates threw their hats into the ring, one of which was John's. His father not only financed John's campaign, but managed it as well. With the wealth and influence of Joseph Sr. behind him, John won 12 percent of the vote—a small amount, but it was enough—more than any other candidate. Still, that was just the primary; he would still have to defeat his Republican rival.[9]

Joseph Sr. plowed $250,000 (approximately $2 million in today's value) into his son's campaign. John campaigned steadily, especially in the working-class areas of his district. His reputation as a war hero, together with the fact his grandfather, Honey Fitz, had held the same congressional office forty years earlier, were all qualities in his favor. However, his youth and inexperience at age twenty-nine could have worked against him. Working with a professional public relations firm, the Kennedys cleverly turned John's youth into an asset with a simple slogan; "The New Generation Offers a Leader" became the campaign's refrain. Joseph Sr. saw to it that the slogan, along with his son's handsome image, was everywhere, bombarding the district with billboards and mass mailings.

Political savvy, deep pockets, a war hero's reputation, all wrapped within a well-run campaign in a heavily Democratic district, made the result an almost foregone conclusion. In the general election, John F. Kennedy easily defeated his Republican rival.

In Congress, Kennedy gained a reputation as a political moderate, and in hindsight many historians place him more in the realm of a conservative than the typical Democratic liberal. Part of this reputation stems from his heavily anti-communist fervor—fervor inherited not just from his father, but earned through his World War II experiences, and those that came after.

During the three terms that Congressman Kennedy served, he was appointed to the House and Labor Committee and the Veterans Affairs Committee—both of which elevated his political visibility. Like so many other key events in John Kennedy's life, both appointments were orchestrated by his father, who continued to ruthlessly leverage his influence on behalf of his family.

Space issues were not on Congressman Kennedy's radar at first. He was more concerned about the nuts and bolts of serving his constituents than he was with grand and glorious visions of satellites and human space travel. But his concerns over the rise in power of the Soviet Union, the worldwide expansion of communism, and a 184-pound orb called Sputnik would force him to alter his focus.

9

SHOWSTOPPER

During an election year no dramatic action may be expected.

—Wernher von Braun

After making a number of design changes, such as increasing the fuel tank size to allow for a 150-second burn time, Wernher von Braun and his fellow engineers in Huntsville knew they had what no one else in America had: a rocket capable of placing a satellite into orbit.

There was just one problem.

In order for the Redstone to reach orbit, the von Braun group had calculated that the booster would need a more powerful fuel. The Redstone booster that von Braun planned to use had the larger propellant tanks required to get into orbit, but the engineers needed something better to fill them with. The oxygen/alcohol propellant combination that had served von Braun so well for over twenty years would no longer be adequate in the soon-to-come, high-performance rocket world of orbital trajectories. The Germans calculated that they would need to boost the performance of their propellants by at least 7 percent in order to push even a small payload into orbit.[1] The von Braun group was made up of the very best rocket engineers in the world, but no one among them could come up with a propellant combination that would not only give them that boost, but would also work with the existing rocket hardware.

This performance deficiency became the bottleneck in America's first satellite launch attempt. Such bottlenecks would later become a part of the jargon of the National Aeronautics and Space Administration (NASA) which referred to these as "showstoppers"—a problem which, if not resolved, would "stop the whole show."[2]

In desperation, the von Braun group awarded a contract to North American Aviation—the builder of the booster—to come up with something that would give them that extra 7 percent of performance. Essentially, the best and the brightest kicked the can down the road.

The contract landed on the desk of the manager of North American's Research Division, Tom Myers. He had hundreds of engineers to choose from to oversee such a project, including a number of very qualified chemists. In the end he decided to give the responsibility to someone who had never headed up an important project before, but who was considered the top theoretical performance specialist in the company. Ironically, that person was not only one of the very few without a college degree, but also the only woman in the company doing engineering work. Her name was Mary Sherman Morgan.[3]

Mary Sherman was born on a farm in rural North Dakota on November 4, 1921, the second youngest of nine children. When she attained school age, her father refused to enroll her based on the belief she would never need an education, and that she was more useful milking cows on the farm. This instilled in Mary a strong desire to go to school. When a social worker intervened, Mary's father said she could not go to school because it was "too dangerous to cross the river to the schoolhouse." The social worker then arranged for the State of North Dakota to give Mary a horse so that she could ride it to school each morning. After threats of legal action from Williams County, the Shermans enrolled Mary in the nearby school—a one-room schoolhouse with eighteen students in eight primary grades.

Mary's teacher introduced her at an early age to the study of chemistry, and by the time Mary reached high school, she had decided she wanted to be a chemist—a career almost completely dominated by men. After two years of college in Ohio, her education was interrupted by World War II and she was recruited to work for a nearby munitions plant, Plum Brook Ordnance Works. Though she never completed her college education, on the strength of her chemistry work at Plum Brook she was hired after the war to work at North American Aviation's new rocket facility in Inglewood, California.

After several years at North American, Mary had established herself as the "go-to" person when it came to the development of new and exotic rocket propellants. And it was this specialized area of expertise that caused Tom Myers to put her in charge of the new Redstone propellant contract.

After several months of work, Mary and two of her colleagues, Bill Webber and Toru Shimizu, came up with a rocket propellant "cocktail"—a mixture of two fuels that would work as one. They told von Braun to keep LOX (liquid oxygen) as his oxidizer, but to replace the Redstone's alcohol fuel with a 60/40 mixture of unsymmetrical dimethyl hydrazine (aka UDMH) and diethylenetriamine (aka DETA).

UDMH by itself would have given the Redstone sufficient specific impulse to get into orbit all on its own, but UDMH had one drawback: its density was too low to be able to fit enough of it into the rocket's fuel tank. DETA alone was not powerful enough to launch the rocket into orbit, but it had the benefit of being extremely dense. Mary Morgan (now married to coworker G. Richard Morgan) came up with the idea of mixing the two propellants together in order to get the best of both worlds—greater power with the UDMH, and a denser fuel with the DETA. The von Braun contract required a 7 percent increase in performance—which many engineers said would not be possible. Yet Mary's cocktail held the possibility of giving the rocket an impressive increase of 10 percent—if one could believe the math.

The first time North American mixed up a batch of Mary's cocktail and ran it through a Redstone engine, the motor experienced combustion instability problems so severe that the test stand's automatic systems shut down the engine only a few seconds after ignition. This was a system that was required by the contract to be able to reliably sustain a 150-second continuous burn. It took several months of head-scratching and trial-and-error testing before North American was able to satisfy all the contract requirements and get the showstopper moving.

When the army asked Mary what she wanted to call her invention, she replied "bagel," on the whimsical notion that the Redstone would be said to be powered by "LOX and bagel." The U.S. Amy did not share her sense of humor, and they ended up calling it hydyne.[4]

With the Redstone's propellant showstopper unstopped, there began to be whispers inside the halls and corridors of the fledging U.S. space program, preaching the belief that the United States would soon put up the world's first man-made satellite.

10

HULLABALOO

A colossal panic was underway, with congressmen and newspapermen leading a huge pack that was baying at the sky.

—Tom Wolfe, *The Right Stuff*

Of the many innovations in rocket technology invented by the Russians, one of the most copied would be their technique of preassembling a rocket in upright flight position inside a large hanger, then slowly tractoring the rocket to its launch pad. NASA would one day adopt this method, but the Soviets found the wisdom in it first. On the morning of October 4, 1957, Sergei Korolev, the Soviet Union's Chief Designer, walked casually beside the massive tractor platform that carried his R-7 rocket and its satellite payload. The trip to the launch area would take approximately one hour, and he wasn't about to let anyone but himself watch over his newborn baby during that hour.

Once the rocket reached the launch area, fueled and ready to fly, Korolev turned over control of the launch, per Soviet custom, to the military. The leader of the launch crew was Colonel Aleksandr Nosov, and he announced the time to launch every few minutes, until finally there were only sixty seconds left. "One minute to go," he announced. "Key to launch."

Lieutenant Boris Chekunov inserted a key into the command console and turned it, arming the launch sequence. "Key on," said Chekunov.

Nosov ordered the engine feed lines to be purged with gaseous nitrogen in order to flush out any residual fuel or oxidizer that might be lingering from the propellant loading process. "Key to drainage," said Nosov, and Chekunov responded by shutting off the liquid oxygen relief valves.

Two minutes were allowed to pass to make certain the lines were fully flushed, then Nosov gave the order: "Pusk!" A single word, meaning "launch."

Chekunov pressed a button that triggered a series of automatic events, like a line of falling dominos. Valves opened, allowing gaseous nitrogen to pressurize both propellant tanks; the umbilical connections were retracted; the rocket was placed on internal battery power; igniters inside the combustion chambers were switched on; the turbopumps were actuated, forcing the propellants into the combustion chambers; the igniters torched the propellants; and a brilliant upside-down volcano of fire poured from the rocket engine nozzles.

"Primary stage!" shouted Nosov as the turbopumps methodically throttled up to full power.

The rocket began a ponderous upward climb.

"Liftoff!"

Future best-selling horror writer Stephen King was ten years old in October 1957. One day he was sitting in a movie theater, mesmerized by a sci-fi film entitled *Earth vs. the Flying Saucers* starring Hugh Marlowe and Joan Taylor. It was about aliens from another planet wanting to take over the Earth, and he was enjoying every minute of it. Suddenly the projector was turned off and the screen turned blank. The house lights came on and the theater manager entered the room. He announced to the crowd that the Soviet Union had just launched the world's first satellite, and that it was named Sputnik.

"At this very moment," said the manager, "it is orbiting the Earth directly above us."

According to Stephen King, the news was "greeted by absolute, tomb-like silence."[1]

On the morning after the Soviet Union launched Sputnik into orbit, the people of the United States woke up in a depressive funk. Americans were proud of their democracy and their perceived superiority derived from that form of government. They were, after all, the country of indoor plumbing, washing machines, and Chevy convertibles. American braggadocio was loud and proud, which made it easy to look down on other countries laboring under less democratic institutions. In nothing did this superiority complex display itself more forcefully than when Americans compared themselves to communist countries. The USSR was looked upon as a backward nation, known more for its subsistence farmers and Siberian gulags than for anything approaching advanced technology. So it was with shock, anger, and a little horror that Americans woke up on the morning of October 5 to discover their foregone impressions of the

USSR, and their assumptions of U.S. superiority, were gravely in question, if not flat-out wrong.

No one was more surprised, however, than William H. Pickering. As the recently installed leader at Pasadena's Jet Propulsion Laboratory (JPL), Pickering was in Washington for a week as part of an international gathering of scientists working on the International Geophysical Year. There was a great deal of anticipation by the delegates, many of whom were expecting either the United States or the USSR to announce a launch date for the world's first satellite attempt. By the time the week drew to a close, however, neither country made any such announcement. On the last Friday before everyone was due to pack their bags and head home, there was a large cocktail party at the Soviet Embassy, to which Pickering, most delegates, and the media had been invited. Walter Sullivan from the *New York Times* took several of the U.S. scientists aside and asked them if anyone at the party had mentioned the launch of the Russian satellite announced earlier that day. Since no one had said a thing, Pickering pigeonholed the lead Russian scientist, Anatoli Blagonravov, and asked what was going on. Blagonravov admitted it was true—the Soviet Union only hours before had successfully launched the world's first satellite. The Americans toasted the Russians, and everyone got drunk.[2]

Pickering left the party angry—angry because, like Wernher von Braun, he knew the Americans could have put up a satellite eighteen to twenty-four months earlier if U.S. politicians had not been so focused on the Vanguard rather than on von Braun's ready-to-go hardware. Pickering returned to Pasadena to find a generally disheartened JPL staff.

Of all those who awoke to the news of Sputnik, no one was more upset than Wernher von Braun. His frustration was palpable. For the rest of his life he would have to live with the knowledge that someone else claimed a record that he could have and should have had. Unbeknownst to von Braun, there were about to be many more frustrating mornings thereafter, all courtesy of Sergei Korolev and his engineers (many of whom von Braun had personally trained).

For the United States as a whole, the clamor was immediate.

It started with the media. It did not take long for the American press to "launch" into controversy, happily feeding the paranoia that seemed to naturally permeate the country. A common story that ran was how both countries had a major launch on October 4: the Soviet Union launched the first satellite, and America launched the pilot episode of *Leave It to Beaver*. For the next few months the American media clomped down hard on the "USSR is way ahead of us" theme. Newspapers, radio, and the new medium of television fanned the flames of public anger over the communist nation's technological superiority. To be fair, these were flames that deserved to be fanned, as America had within it the finest rocket scientists in the world and, were it not for politics and a space policy built on waffling, could easily have put up a satellite two years before the Russians.

Though the Russians had superiority over the Americans in rockets, the United States soon proved it was way ahead of the Soviet Union when it came to capitalistic opportunism. Business owners understood the problem: if the Soviets could launch an object around the world, then it could certainly send an atom bomb to America's heartland. And so billboards and newspaper advertisements from construction companies suddenly appeared everywhere, offering their services in the construction of "backyard bomb shelters." Manuals on how to build such shelters became best sellers.

What made the communist satellite so frustrating to American politicians and policy wonks was its benign nature, coupled with its very public profile. It wasn't a weapon—it was just a simple ball of scientific experiments packed with a radio—a radio that sent out an unclassified, unencoded *beep-beep-beep* signal that anyone with a shortwave radio could listen to. And at certain times of the evening the satellite could even be seen with the naked eye, flying silently overhead. It was a nonthreatening public spectacle—a community experience that all humanity could participate in with little effort. And yet it put a popular president in the hot seat.

President Dwight D. Eisenhower is often credited with coining the term "Space Race," but there is probably no way to know for certain who came up with it first. Regardless of its origin, the media latched onto it immediately—the sound-bite catchiness of its rhyme was too good to resist. The term, however, is a little misleading. What we refer to as the Space Race was really nothing more than the latest technology race in a long line of technology races spawned by humans for thousands of years. Several European countries were able to dominate the takeover and colonization of other peoples and places due to winning the technology races of shipbuilding, gunpowder, rifles, field artillery, and more.

The Space Race was just another in a long line of such human contests to discover and grab new technologies. And as had happened so often before, both the United States and the Soviet Union immediately understood that their very survival might be in jeopardy if they didn't win the race, or at least catch up to its leader. Technology races, more than anything else, had spawned the need for covert spying for centuries, and both countries began to ramp up their spy games.

CHAPTER 10

The public going goofy over Sputnik, the media fanning their goofiness, the military demanding more money for pet rocket projects and spy planes, politicians screaming across the aisles: this is where America found itself in October 1957.

Unless you lived in the Mojave Desert.

At Edwards Air Force Base in the dry, sandy desert of California, test pilots whose fame was still in the future such as Chuck Yaeger, Slick Goodman, Iven Kincheloe, Joe Walker, Scott Crossfield, and many others had been testing rocket-powered aircraft for years. In October 1947, ten years before Sputnik, Chuck Yeager had been the first to break the whimsically named "sound barrier" in the rocket-powered Bell X-1. Not only were manned rocket-powered flying machines old hat to the pilots at Edwards, they were just waiting for the next generation of rocket planes to come off the assembly lines—rocket planes destined for human space travel. The most anticipated of this new generation of rocket spaceplanes was the Walter Dornberger–inspired X-15—a machine designed by North American Aviation, whose purpose was to carry a man more than fifty miles above the Earth, the official demarcation of where space began. Once the X-15 was available, man-powered flight into space would become routine—almost blasé. In addition, the pilots had also been made aware that the X-15 was merely the precursor to a plane far more ambitious, the X-15B, which was already in the planning stages. Coupled with a Titan main stage, the X-15B would be able to launch two men into space, orbit the Earth a few times, then allow the pilots to return to Earth in a standard plane-like, wheels-down landing. And this was decades before the Space Shuttle. All of this and more had been talked about, planned, and at least partially engineered long before Sputnik. But all the programs at Edwards were not only top secret—ensconced as they were in a quiet, remote desert—they were also out of sight and out of mind. Few people outside the aviator clique at Pancho Barnes's Happy Bottom Riding Club knew what was going on. What would soon become a major liability was the fact that the military brass who controlled the projects at Edwards, and the pilots that flew them, were ignorant of political realities. Each morning they woke up and went about their daily business of "pushing the envelope," caring little about whether it might be a good idea for American taxpayers and voters to know a few details of what they were accomplishing. All that success bathed in all that secrecy created the perfect breeding ground for the post-Sputnik uproar and head-scratching.

So when America went goofy over the Soviet orb, Yeager, Goodman, Crossfield, and the other pilots had one big question. The little Sputnik satellite weighed 184 pounds—the X-15B would weigh 15 *tons*. What the hell was all the hullabaloo about?

Two thousand miles away in Huntsville, von Braun was well aware of what the hullabaloo was about. Sputnik changed everything for everybody, even if nobody knew it. Wernher redoubled his efforts to put a satellite into orbit, while in Pasadena, Pickering was strategizing how to leapfrog the Russians by launching a robotic craft to land on the Moon. He called it "Operation Red Socks," and everyone he presented the idea to agreed it would be a viable response to Sputnik. Everyone, that is, until the proposal got as far as Washington, at which point the Pentagon put the kibosh on it. Like President Eisenhower, the military brass still favored the Vanguard over the Redstone. Emboldened by his new TV celebrity status, von Braun sang the praises of his Juno rocket (the "civilianized" version of the Jupiter ICBM) everywhere he went—especially wherever and whenever the media assembled. Eisenhower eschewed the very idea of competition in space and was angered by what he perceived as von Braun's self-serving trumpet. "When military people begin to talk about this matter and to assert that other missiles could have been used to launch a U.S. satellite sooner," said the president a short while after the Sputnik launch, "they tend to make the matter look like a race, which is exactly the wrong impression."[3] Eisenhower may have been the most powerful man on Earth, but he was blind to how that Earth was changing.

After having his enthusiasm and his Red Socks program deflated, Pickering decided it was time to put his eggs in the right basket. Partnering with James van Allen and von Braun, Pickering and JPL designed a final-stage rocket and satellite to put atop the Jupiter C booster. They essentially planned for failure—failure of the Vanguard. All three men were so convinced the Vanguard would fail in its mission that they put in place a series of decisions that would make their rocket the unofficial Vanguard backup—a backdoor plan that would succeed in spectacular fashion.

On December 6, 1957, the U.S. Navy mounted its latest Vanguard rocket on a launch pad at Cape Canaveral. As the count reached zero, and the launch sequence started, the rocket rose a few feet into the air, lost thrust, then collapsed backward, exploding under the weight of its topped-off propellant tanks. With hundreds of media representatives watching, the rocket's mighty demise made the front page of every newspaper, the cover of a hundred magazines, and the six o'clock news for weeks thereafter.

Destiny oftentimes comes not from fate, but from preparation, and the Van Allen/Pickering/von Braun triumvirate was ready and prepared. Like an understudy waiting in the wings for their big opportunity, they stepped forward. At a Washington party soon after the Vanguard explosion, von Braun gave a warning to an Eisenhower cabinet member, declaring, "The Vanguard rocket will not make it!" He went on to boast that he and his team "could have a satellite up in sixty days."[4] Given the political heat everyone was under, it was a boast that required attention.

Eisenhower—author of the now-famous warnings regarding the "military-industrial complex"—misunderstood his constituency, and he certainly misunderstood von Braun. He considered von Braun's speechmaking and drumbeating to be a symptom of that military-industrial alliance monster he so dreaded. But though von Braun's official alliance was with the U.S. Army, his true allegiance was elsewhere—to that twelve-year-old Lutheran boy who first looked through his telescope and began to dream of building his own rocket to fly to the Moon. Eisenhower could not understand that one man with boyish dreams could so move public opinion and policy. Over the next few years, many American politicians would learn a great deal from the German expatriate.

Eisenhower's lack of enthusiasm for the space program, orbital satellites, or what the Germans could bring to the table of American technology and space achievement were revealed on the evening of the successful orbital launch of America's first satellite, Explorer 1. Just after midnight James Hagerty, the White House press secretary, called the president with the news. After remarking that it was a wonderful achievement, the president said, "Let's not make too big a hullaballoo over this." When Eisenhower made the official White House announcement, it was the epitome of downplay: "The United States has successfully placed a scientific Earth satellite in orbit around the Earth."[5] The announcement noted that the satellite was part of U.S. participation in the International Geophysical Year activities, but did not proclaim its historical significance. And it made no mention of any space race.

Rank-and-file Americans, however, felt quite different. When the papers hit the doorsteps the next morning, and the news was revealed on radio and morning television shows, the country celebrated. They knew the Space Race existed even if Washington did not, and they exulted in finally showing the rest of the world that Americans did not intend to "wallow in the backwater of space," as John F. Kennedy would so eloquently promise a few years later.

Almost immediately Wernher von Braun began to light a rocket-hot fire under the "hullaballoo" that Eisenhower wanted to avoid. He did so using three methods. First, he told everyone who would listen how great a success he and his engineers had given the United States with the Juno rocket. Second, he continued to appeal to the public's imagination about the future wonders of space travel. And third, he shamelessly used fear to motivate those who were unmotivated by the first two methods. And he made sure that fear spread to the U.S. Congress. At a press conference, when asked if the launch of Sputnik meant that the Russians now had the power to hit Washington with a hydrogen bomb, he replied, "Yes."[6] Von Braun could be dramatic when he needed to be, but he had an immense talent for humor as well. Soon after the Soviets launched Sputnik II with a live dog aboard—indicating they were working on life-support systems for space—a reporter asked von Braun, "What do you think American astronauts will find when they land on the Moon?" Wernher replied, "Russians."[7]

Eisenhower's space myopia seemed to have no bounds. Once the United States had its first satellite up, it finally gained some bragging rights that until then had been the sole possession of the Soviets. The country and its leaders should have crowed about their success and shouted it from the rooftops, yet the president wanted the exact opposite. Having been bred within the military, he had a military mind-set and was not the best of politicians. When Eisenhower heard of Sputnik II and the spacefaring dog Laika, he was frustrated at how everyone else seemed to think it was an important step. He eschewed the whole idea of civilians enjoying space through discovery and adventure—he wanted the focus on weapons and defense, and that meant ballistic missiles. The adventurous space program envisioned by von Braun, showcased by Disney, and quickly embraced by the American populace was, in the president's mind, a waste of money. Why go to the Moon, he asked, if we don't have any enemies there? This attitude was reinforced when it was discovered in the late 1960s that less than a year after the launch of Sputnik, Eisenhower had secretly put his weight behind what would become the world's first spy satellite, later euphemistically named GRAB (Galactic Radiation and Background).

When we think of the Saturn V Moon rocket today, we usually connect it in our minds with President Kennedy. What is not well known is that President Eisenhower had the proposal for this project on his desk prior to Kennedy's 1960 election victory. NASA quietly presented to the president its plan to eventually land a man on the Moon—a goal that worried Eisenhower. He had his science advisor, George Kistiakowsky, form a six-man committee to study the proposal. The committee's findings were laid on the president's desk one month prior to Kennedy's inauguration. The projected cost was astounding at the time: $8 billion. When Eisenhower discussed the proposal with his advisors, one of them noted, "This won't satisfy [them]. When they finish this, they'll want to go to the planets." The comment generated a great deal of laughter in the room.[8]

Two months after Explorer 1 sailed into orbit, the poor, pitied, oft-lampooned Vanguard rocket would launch America's second satellite, Vanguard 1, into orbit. It would score a series of impressive successes, including becoming the world's first solar-powered satellite. Today Vanguard 1 holds the record for the largest number of Earth orbits of any man-made object. While Explorer 1 has long since burned up in Earth's atmosphere, Vanguard 1's orbit is so solid that NASA has calculated it will continue to circle the Earth for at least another ten centuries. Yet Sergei Korolev's predictions about the inevitable forgettability of every "number two" has turned out to be prophetic; today almost no one remembers, or has ever heard of, Vanguard 1.

11

THE EARLY PROBES

There are nights when the wolves are silent and only the Moon howls.

—George Carlin

Given the Soviet's huge lead in space in the late 1950s and early 1960s, it is easy to believe the historical rhetoric that the Russians were giving Sergei Korolev everything he wanted, while U.S. scientists were being starved for resources. That scenario was more fable than fact. Korolev wanted to put a man in orbit as early as 1957, and probably could have if the Soviet powerbrokers had given him the money, manpower, and equipment he needed. In reality, the Soviet government, like the U.S. government, was so focused on weapons and ICBMs that they could see no point in building manned spacecraft. In 1958 the Soviets finally authorized Korolev to begin work on a manned space capsule, but budgeted no money to actually send it aloft. The Soviet military wanted satellites that could pass over the United States and other countries and point spy cameras below to keep watch. Khrushchev and the Politburo had feelings similar to Eisenhower—cosmonauts flying around in space just seemed like so much joyriding. The message given to Korolev was blunt: a "reconnaissance satellite is more important for the Motherland."[1]

To make matters worse, Korolev eventually lost his direct access to Khrushchev, and was forced to go through intermediaries in order to procure equipment and supplies. This only slowed production further, and greatly frustrated the Chief Designer.

Denied funds for a manned spacecraft, Korolev took a different tack. While Americans were patting themselves on the back for finally getting into orbit with Explorer 1, the Soviets were making plans to launch the first space vessel to leave Earth orbit. Korolev had ordered his engineers to begin design work on lunar probes—unmanned robotic craft that could orbit, or even land on, the Moon. There were pragmatic reasons for this temporary change in emphasis. If the Soviets were going to send cosmonauts to the Moon, it was urgent that they have some idea what the landing field was like. Was the lunar ground hard-packed, or was it so soft and powdery that a landing craft would sink? Robotic probes held the promise of offering up an answer.

Korolev, von Braun, and pretty much anyone who had ever worked in the space business understood the advantages of probes over manned spaceflight. For one, there is an approximate ten-to-one, results-to-budget ratio; that is, you get ten times more data for every dollar spent on a probe mission than on a manned spaceflight. Not having to include life support systems and re-entry devices saves a great deal of time, weight, and money. Unmanned probes can go where humans cannot, such as close approaches to the Sun, or extreme distances to the outer planets. Even if the space scientists of the 1950s, or even today, were given huge budgets, probes will always have certain advantages over manned spaceflight.

Korolev was aware of the advantages of probes from the very beginning, and always had on his draft board ideas for such spacecraft. One of Korolev's goals was to see what the back side of the Moon looked like. For millions of years tidal forces had locked the Earth/Moon system into a peculiar dance whereby one side of the Moon was always facing toward the Earth, no matter their orbits, no matter their rotation. No human being had ever seen what was on the other side. Powerful telescopes could tell us what all sides of Mars, Saturn, and Jupiter looked like, yet the far side of our nearest neighbor remained a complete unknown. What was back there? Korolev was determined to find out, and so he began to plan a series of daring robotic lunar flybys and landings.

However, despite the Soviet's space lead, the first attempt to send a probe to the Moon was made by the Americans. Using a Thor-Able rocket, the U.S. Air Force had a lunar probe, dubbed Pioneer 1,[2] sitting atop their rocket on August 17, 1958—a mere seven months after putting up its first satellite, and two months before NASA had even been established.[3] But when the air force controllers pressed the button to launch, the Thor-Able rocket went nowhere, failing to even lift off the launch pad.[4]

NASA and the Soviets each made three more attempts to launch lunar probes, and every one of them failed for various reasons, mostly due to propulsion problems. Both countries were discovering the many drawbacks of science as a race rather than a voyage of discovery. Too much concern for speedy progress coupled with too little meticulous preparation were causing a "hurry up and hope" mind-set. The result would be a long string of failures by both countries.

It wasn't until January 4, 1959, that the USSR finally was able to claim some level of success. A Korolev rocket carried a payload dubbed Luna 1 toward the Moon, intending it as an impactor.[5] Luna 1 missed its target by six thousand kilometers, speeding past the Moon and settling into a solar orbit, where it remains to this day. Despite this failure, it was the first time any human-made machine had left Earth orbit, and in that regard it was still a major success.

Illustrating how young and callow the technology was, the Luna 1 flyby laid claim to being the first success out of eight attempts between the two countries, and even then the mission's most important goal was not achieved.

Another Soviet lunar probe failure occurred six months later. Then, in March 1959, the United States scored its first lunar probe success. Pioneer 4 performed a flyby, coming to within thirty-seven thousand miles of the Moon's surface, before entering a solar orbit. Parts of its instrument package failed, which is why it is listed in NASA records as a "partial success."

Once the sole owner of first place in the Space Race, the Soviets found themselves in something more akin to a leapfrog contest; they would score a space first, then the Americans, then the Soviets again, and so forth. The USSR once again took the lead on September 14, 1959, when Luna 2 successfully impacted on the Moon—the first man-made object to "land" there. The United States followed up with another lunar probe attempt, which ended with the explosion of its booster rocket on the pad.

The first real major success of a lunar probe occurred on October 7, 1959, when the Soviets successfully placed Luna 3 in orbit around the Moon and were able to have it take photographs of its far side. The back side of the Moon would no longer be a mystery—but only if the photos could be transmitted to Earth, received, and displayed.

In 1959 all photography was based on chemical emulsion—digital photography was still a future invention. Therefore, for Luna 3 to transmit photos back to Earth by radio, it first had to develop the photos remotely, on board the craft. The images would then be sent via a radio analog signal to Earth.

Soon after, however, Korolev received word from the radio station in the Crimea that they were having trouble receiving the photos. Always a hands-on individual, Korolev grabbed a number of engineers and senior staff members and flew to the Crimea. He had arranged for a helicopter to take them to the radio station, but bad weather forced the seven men to hire a car and driver instead. After a hair-raising ride in a snowstorm, Korolev and his entourage arrived at the station. The Crimean technicians were more than surprised when the Chief Designer himself and his entourage barged through the door. Given that Korolev's identity had been such a well-kept secret, the radio station technicians had no idea who he was at first. The rocket men wasted no time in taking over and immediately began their investigation. Korolev and his men soon figured out that the station operators were not performing their task properly, and before long the photos were being received.

The Luna 3 mission went down in the history books as the most important and spectacular space success up to that point. Eighteen photographs of the far side of the Moon, covering 70 percent of its surface, were eventually received and published. Following the long-standing tradition of human exploration, Korolev began assigning names to the mountains and other geographic landmarks shown in the photos. To this day most of the far side craters, rills, and valleys have Russian names, and probably always will.

Luna 3 was a spectacular success, but luck as much as anything played into it. Twelve of the next thirteen lunar probes launched between the two countries would end in failure. Starting in August 1961, NASA and JPL launched a series of impactors called the Ranger program. The goal was to stream hundreds of photographs back to Earth as the Ranger probes approached the Moon's surface, up until the moment of impact. The first six Ranger probes failed, causing Congress to intervene and make some management changes within the two organizations. Finally, on July 28, 1964, NASA launched Ranger 7, which scored a direct hit in the Moon's newly named Mare Cognitum, sending back more than 4,300 photos from six cameras. It was followed seven months later by Ranger 8, impacting in the Mare Tranquillitatis, and Ranger 9 a month after that, hitting Alphonsus Crater. The last two Ranger missions would prove crucial, as they impacted in areas under early consideration as potential manned landing sites. As a result of the data sent back by Rangers 8 and 9, NASA scientists concluded that landing a manned spacecraft on the Moon would probably be safe (i.e., it wouldn't sink into some deep quicksand-like lunar powder). Korolev's prophetic declaration was, tentatively, verified.

It would not be until June 1966, with a more advanced probe—Surveyor 1—that this probability became more of a certainty.

12

THE MERCURY PROJECT

It's a very sobering feeling to be up in space and realize that one's safety factor was determined by the lowest bidder on a government contract.

—Alan Shepard

In February 1959, military test pilots around the country began receiving blind orders—orders to travel somewhere without being given a reason.[1] The orders summoned them to a meeting at the Pentagon. Upon arriving, each discovered they were part of a small group of about thirty-five attendees, all of them test pilots, all of them officers, from all branches of the military. In the meeting they were told they were the first candidates in a new manned space program, Project Mercury, and that if chosen they would be taking a leave of absence from their military careers.

The NASA brass had originally intended to have an "open call" for astronaut candidates, that is, almost anyone who met a set of height, weight, age, and education standards could apply. But when President Eisenhower heard of it, he overturned the decision, insisting instead that the first criteria for a candidate was that they be an experienced test pilot. NASA placed an advertisement in a journal read by military test pilots and received more than five hundred applications. Until the day some of them received their blind orders and showed up at the Pentagon, none of them had a heard a word about the status of those applications.

One of those in the initial call-up was Wally Schirra. In late 1958 he was stationed at the Test Pilot School (TPS) Patuxent, a U.S. Navy flying school near

Annapolis, as a test pilot for the F4D, a Mach 1 fighter jet. By this time there were plenty of Mach 2 aircraft coming off the production line, and so the F4D was considered a "slow horse" in test pilot circles. Still, it was Schirra's job to learn how to fly it, run it through its paces, and write up reports.

One evening, shortly before graduation, Wally was writing one of those F4D reports when a fellow pilot called him outside and pointed to the sky. The two men gazed upward as a pinprick of light slowly made its way across the heavens. It was dusk, and the sun was in just the right position to highlight the gleaming metal exterior of Sputnik—the Russian satellite launched the previous year.

The moment turned into an epiphany for Schirra. "I wondered why I was messing around with an airplane that could barely make Mach 1, when a Russian satellite did Mach 25. It was one of the few times until then that I had given any thought to going into space myself."[2]

As the first morning meeting closed for a lunch break, many of the candidates congregated in groups, mostly by their military affiliation, and discussed their new situation. All of them had concrete plans for how the arc of their future military careers was going to go, and taking a leave of absence for NASA (which could take five to ten years off that career arc) was not on any of their lists. Some were reluctant and refused to participate, but many of them decided to try out. Anyone who didn't make the cut, after all, would simply resume their military career where they left off. Those who did decide to sign on had pretty much the same motivation: for a test pilot, there was no greater need than to put themselves into the history books by being the "first" at something. They reveled in it, bragged about it: Chuck Yaeger (first man to fly supersonic), Scott Crossfield (first man to fly Mach 2), John Glenn (first man to fly supersonic from coast to coast), and so forth. According to Schirra, "Test pilots of high-performance aircraft collect trophies like some people collect stamps."[3] But all the best jet aircraft records (and trophies) had been taken, and new "firsts" were becoming much harder to come by. NASA offered something they could no longer look forward to: a very long list of new firsts just waiting to be plucked from the glory tree.

With great reluctance and trepidation (as regarded their careers, not the danger), twenty-four of the original group of thirty-five candidates who had been summoned to the initial Pentagon meeting ended up signing on to try out for a spot on the astronaut roster. About another two dozen signed up from a second summoned group.

The astronauts were being recruited for Project Mercury. NASA's master plan for Mercury called for seven space flights, each with a single astronaut on board. Seven flights, seven astronauts. The first three flights would be "suborbital"—a euphemism that simply meant a ballistic flight: the rocket would go up into space, then return to Earth by simple gravity, like throwing a baseball high into the air. Due to some behind-the-scenes machinations by John Glenn, the plan changed, and only the first two flights were ballistic, with the third flight an orbital mission (manned by Glenn, of course). When the flight plan was proposed to President Eisenhower, one of his aides, George Kistiakowsky, was highly critical of the entire program, predicting it would be the "most expensive funeral a man ever had."[4]

The original astronaut selection plan had been to choose twelve prospects from all the candidates on the theory that the testing and training program they would endure would be so rigorous, some would drop out. But the decision to use only test pilots made that assumption moot—it quickly became obvious that these men were not the kind to drop out of anything.[5] So instead of choosing twelve recruits and winnowing by attrition down to six (later revised to seven), NASA decided to simply choose the final flight group with no alternates. If something didn't work out, they could always go back to their talent pool, or have one astronaut fly twice.

The most rigorous mind and body examination program ever conducted began, with physical testing in New Mexico, and psychological testing in Ohio. Never before had a group of human beings been so studied, examined, poked, prodded, and peered into as what these men were about to experience. Despite the rigor, paring down the astronaut pool became a larger-than-expected challenge when almost every candidate passed his tests. These were men who did not suffer failure well, and dropping out was just not in their DNA. Even so, one by one the system began to winnow out a few, and the cream floated to the top. When all was said and done, NASA chose its seven: John Glenn from the U.S. Marines, Walter (Wally) Schirra, Alan Shepard, and Scott Carpenter from the U.S. Navy, and Donald (Deke) Slayton, Leroy Gordon Cooper, and Virgil (Gus) Grissom from the U.S. Air Force. The agency's decision, however, remained secret. NASA was becoming more publicity-savvy, and it developed plans for a major press conference during which it would announce the final astronaut selections.

By the summer of 1958, Thomas Keith Glennan had enjoyed a career that could best be described as unusual. A graduate of the University of Wisconsin, Eau Claire, with a degree in electrical engineering, Glennan worked his way up the ranks of the film industry by utterly transforming it, bringing sound technology into what had been a silent medium. He eventually became the head of two Hollywood studios, Paramount and Samuel Goldwyn. World War II thrust him into work for several institutions, including the U.S. Navy Underwater Sound

Laboratory. An executive at Ansco Corporation, president of Case Institute of Technology, and member of the Atomic Energy Commission all eventually ended up on his eclectic résumé. Then, on August 19, 1958, he was appointed NASA's first administrator.

A mere eight months later, he was informed by the astronaut selection committee who the chosen finalists were. He scheduled the press conference for April 9, 1959.

On the appointed day, journalists and photographers from all over the world descended on Washington for the announcement. Glennan entered and stood before them, officially introducing the media and the world to the seven men chosen for the U.S. Mercury missions into space. They were to be called "astronauts," he explained, after the pioneers of ballooning who had been called argonauts.[6] In front of the cameras, smiling, and dressed in civilian clothing, the seven future astronauts appeared without their engineering/military officer/test pilot identities in evidence. Their military characteristics had been covered over in order to form a more approachable public persona—essentially stripping them of who they really were.[7]

The astronauts were seated on a riser so that the large throng of journalists and photographers could see them. Left to right, the men were placed alphabetically, leaving John Glenn sitting in the center. This fluke of positioning turned the meeting into an amazing piece of theater. After Glennan introduced each astronaut by name, the reporters were allowed to ask questions. It was then that NASA and its astronaut corps would discover they still had a few things to learn about public relations. The astronauts had been focused on qualifying for the Mercury program. They were now ready to concentrate on their upcoming training, their mission assignments, the science and adventure of flying into space. These were the subjects they were ready and prepared to answer questions about. But the reporters wanted nothing of it. To the surprise of the astronauts, the questions from the reporters were of the "hearth and home" variety. The room of journalists wanted to know about wives, children, family, duty, and religion. While most of the astronauts (some of whom had very little in the way of bragworthy family lives) froze like deer in the headlights, John Glenn answered openly and with enthusiasm. For Glenn, the only Marine in the group, these types of questions were fertile ground. Mom and apple pie, America and God—Glenn suddenly found himself in his element.

When a reporter asked about whether their wives were supporting them in their dangerous new endeavor as astronauts, Glenn was ready: "I don't think any of us could really go on with something like this if we didn't have pretty good backing at home, really. My wife's attitude toward this has been the same as it has been all along through my flying. If it is what I want to do, she is behind it, and the kids are, too, a hundred percent."[8]

In his book, *The Right Stuff,* Tom Wolfe eloquently describes the dilemma the other astronauts faced with first having to deal with the unexpected questions, and then having to deal with John Glenn.

> What the hell was he talking about? *I don't think any of us could really go on with something like this. . . . What* possible difference could a *wife's attitude* make about an opportunity for a giant step up the great ziggurat? What was with this guy? It kept on in that fashion. Some reporter gets up and asks them all to tell about their religious affiliation (*religious* affiliation?)—and Glenn tees off again.[9]

They were trained in the operation of planes and jets. They understood flight plans, how to bank and when to yaw. They knew how to eject and how to parachute safely to Earth. They were trained in water rescue and desert survival. They understood switches, buttons, and meters. But John Glenn's performance at their public unveiling showed all of NASA and the astronaut corps a key skill they had no training or experience in whatsoever: charm.

While the nation's attention was riveted on the human drama of astronaut selection and glorification, Wernher von Braun and his engineers and technicians at the Army Ballistic Missile Agency in Huntsville, Alabama, as well as the engineers and technicians at the Convair Division of General Dynamics in San Diego, California, were fast at work converting their respective Jupiter C and Atlas rockets from ICBMs into man-rated flying machines. For von Braun this was a distraction as he, ever the forward thinker, was already on task designing what would become the heavy-lifting Saturn class of rockets that would one day be adapted for use in the Apollo lunar landings.[10]

If man-rating the Jupiter C were not distraction enough, Wernher found himself frequently mired in politics. With the United States having a growing and emerging rocket science field, von Braun's German engineers were in demand—offers of employment from industry were flowing in. And those offers were consistently better than what the army was paying. Von Braun already had experience with competition for his resources. During the war, Himmler's SS had tried to fold the Peenemünde group into their organization. To prevent such a move, von Braun arranged with the Speer ministry to convert his entire development group into a government corporation. Now, with American companies competing for his people, and NASA pressing to have the Huntsville group folded into the camp, Wernher had a similar dilemma on his hands. He wrote a confidential memorandum to General John Medaris, offering several recommendations. In it he illustrated the problem: "We have the very imminent danger of loss of a number of our key scientists and engineers" due to the "highly attractive and lucrative offers from industry."[11] Von Braun suggested to Medaris that the best solution might be to detach the Huntsville development group from the army and convert it to a private company. There is no record that Medaris ever responded to the memo, but it is known that behind the scenes, von Braun had sent representatives to California in an attempt to have either Raytheon or Solar Aircraft take over their operations. Whether for pragmatism, nostalgia, or both, von Braun was desperate to keep his team of German scientists together as a cohesive engineering unit. His best attempts, however, would fail. Over the next few years, his scientists and friends were picked off one by one, lured into higher-paying jobs in private industry. Even his close friend and confidant Dieter Huzel[12] would end up in California, helping to design the powerful F-1 rocket engine for North American Aviation. Five F-1 engines would later be incorporated into the Saturn rocket, so in a way, Wernher and Dieter were still working on the same team.

Despite the many political and financial distractions, von Braun was determined to show everyone who the best rocket scientists in the world were. Once the call came from NASA that his Redstone/Jupiter C booster would fly the first two U.S. astronauts into space, he was determined to make his rocket as perfect and flawless as possible. He insisted on a number of design changes in the booster to increase its reliability—changes that would end up delaying Alan Shepard's first launch. He and his men then had to endure complaints of "German over-engineering."[13] Those complaints, however, were muted by the fact that the construction of the Mercury capsule by McDonnell Aircraft was delayed due to unforeseen complexities with the emergency escape system.

While rocket hardware was being designed and tested, and the new astronaut corps was going through its training and preparation, another group of NASA managers and scientists were quietly working on a less glamorous, but no less important, issue: How do you make a spacesuit?[14] Thanks to the military's experience with high-altitude jet fighters, a great deal was already known about life-support systems in low-pressure, supercold environments. This experience would come in handy now that a garment had to be created that would keep a human healthy and happy in an ultracold, zero-gravity vacuum. And, in fact, when NASA appealed to contractors for proposals for the suits, they made it known they were looking for something like the navy's Mark IV flight suit, only better. Initially, NASA astronauts would not require fully self-contained spacesuits like those that would be needed a few years later when astronauts began spending many hours and days in orbit or leaving their capsules to perform EVAs (extravehicular activities). For Mercury, something less complicated was all that was needed.

NASA received bids from the David Clark Company, B. F. Goodrich, and, as odd as it may seem, the bra manufacturer Playtex. Goodrich's bid ended up winning the contract. Not to be outdone, however, David Clark would later win the contract to make pressure suits for the Gemini program. Playtex would never make an astronaut pressure suit, and perhaps the alpha-male astronauts preferred it that way.

The Mercury 7 astronauts soon settled into their new homes in Cocoa Beach, a small community fifteen miles south of Cape Canaveral. Though the astronauts had become international celebrities due to their new career path, locally they were famous (or infamous, depending on whom you talked to) for street racing their sports cars. As a public relations gimmick, General Motors had loaned each of them a white Corvette for a one-year period, what were then termed "brass hat cars"—cars for VIPs that could be sold later for a premium due to who had driven them—in this case, NASA astronauts. After the year was up, the astronaut could either return the car to General Motors or purchase it outright. Either way, the company received millions of dollars in free publicity.[15]

The problem with loaning seven fast sports cars to a bunch of jet fighter jocks, however, soon became apparent as street racing among the astronauts became all too common, occasionally resulting in accidents and brushes with the law. Alan Shepard, in fact, was once caught exceeding the speed limit so severely that he was arrested. In another Alan Shepard incident, the astronaut was driving a borrowed Ford Shelby GT with Gus Grissom and Wally Schirra along as passengers. When the light turned green at an intersection, Shepard decided he wanted to see what kind of acceleration he could get out of the car and floored the accelerator. Unfortunately, the trio had failed to notice the police car that had been stopped beside them at the light. As the Shelby tore off, tires screaming, the police car took off in pursuit.

Seeing the lights of the local law in his rearview mirror, Shepard pulled the car over and got out, waiting for the officer to approach. The two officers immediately recognized him and started laughing. "You're Al Shepard, aren't you?"

"That's right."

"And who's that in the car? Is that Wally Schirra?"

"Yes sir."

The officer admonished Shepard to obey the speed limit, and then let them go without a ticket. It was one of many times that being a famous astronaut would get them off some hook or deliver some benefit. Free drinks and dinners could be found up and down the Florida coast, and pretty much everywhere else. A few years later, as the launch of Gemini 6 approached, a local tire dealer offered Schirra a free set of tires for his car if he pulled off a key goal of the mission: a rendezvous with a second orbiting vessel. Such offers were common, and made the job of astronaut far more lucrative than anyone had anticipated. NASA attempted to keep a lid on celebrity endorsements and other business deals that would exploit the astronauts, but it was impossible to watch them all the time. At one point the astronauts even became mutual business partners, each putting up $10,000 to purchase shares in a brand-new Florida hotel. The hotel would eventually fail, but after everyone sold their shares to an investor, they still made a tidy profit, literally doubling their investment.[16] Even in failure, the Mercury 7 astronauts would usually find ways to succeed.[17]

In early 1960 a caravan of Lincolns and Cadillacs pulled into the small coal-mining town of Welch, West Virginia—population 1,500. In a poverty-laced place where most cars were rusted and dinged up, these high-end vehicles stood out. An advance team had placed posters all over town in preparation for this visit—a visit from a man who was running for president of the United States.

Walking along the streets of Welch was a young boy who was visiting from Coalwood, another small coal-mining town nearby. The boy had beaten incredible odds, winning the state science fair with a project so fascinating he had been invited to compete at the National Science Fair in Indianapolis. No one from coal country had ever achieved this, and everyone was proud. The boy's mother wanted to make sure he looked presentable for the event, so she had sent a family friend, Emily Sue, to drive him to Welch and buy a nice suit (which the boy absolutely did not want). When the young woman wasn't looking, the boy, out of spite, picked out a bright orange suit, put it on, and paid for it. As he exited the store, Emily Sue was aghast at what he had purchased. Her vocal reaction was interrupted, however, by the Lincolns and Cadillacs pulling up and stopping close to where they were standing. Men got out of the cars and hoisted a skinny man with wavy hair on top of one of the Lincolns. The boy knew right away who he was.

The senator began to speak, but the small assemblage of Welch's citizenry that assembled did not appear interested at all. The politician tried hard, though, doing his best to rile up a people that had no intention of being riled. Finally, exasperated by the doldrum crowd, the man asked if there were any questions. The boy in the orange suit raised his hand high, and the senator called on him.

"Yessir," said the boy. "What do you think the United States ought to do in space?"

The senator threw the question back. "What do *you* think we should do in space?"

"We should go to the Moon!" said the boy.

"And why do you think we should go to the Moon?" asked the senator.

"We should go there and find out what it's made of and mine it just like we mine coal here in West Virginia!"

The boy did not care much for coal mining, but he knew his comments would loosen up the crowd, and that is exactly what happened. What could have been a boring whistle-stop on the senator's campaign ended up being enjoyable and successful as the crowd began to engage with the man. Twenty minutes later, when the candidate rode off with his entourage, many were sad to see him go.

Senator Kennedy would win that presidency, and he would one day take that boy up on his advice. The boy's name was Homer Hickam, and he would write a book called *The Rocket Boys*, which would be turned into the film *October Sky*. He would also grow up to be a rocket engineer and help bring to fruition not only his boyhood dream, but a president's audacious vision.[18]

13

MISSION: CONTROL

In the pre-Apollo days . . . there was some early resentment. It was not the most harmonious of relationships.

—Frank Williams, assistant to Wernher von Braun

We knew nothing about rocketry, we knew nothing about spacecraft, we knew nothing about orbits.

—Gene Kranz

Once the U.S. space program began to coalesce into a reality involving plans for actual manned spaceflight at a cost of many millions of dollars, another competition began, and not just a competition for resources and money. It was a competition over philosophy and the very direction of the country's space program. Man had never flown in space before, and so there was no rule book, no long-established traditions, no manual of procedure. Most significantly, the executive hierarchies were so often in flux that there seemed to be no one in charge. Various aspects of the country's nascent space program were squabbling and acting like fiefdoms, refusing to share what they were working on with others.

A prime example of this "fiefdom mind-set" occurred when Wernher von Braun attended a meeting at Langley Research Center in Virginia, where engineers and department heads had gathered to discuss the upcoming Mercury program. Wernher had gotten wind of a two-man capsule that was on the drawing boards (the precursor to the Gemini spacecraft). One of the leaders of the meeting was Robert Gilruth of the National Advisory Committee for Aeronautics (NACA).[1] Von Braun wanted more information about the rumored two-man capsule, and when he asked, "Would you care to give us a briefing on it?" Gilruth replied, "No, I wouldn't."

Of all the country's space enthusiasts, the only one with a clear vision was von Braun, but he was far down the ladder of authority, and his vision was often too grandiose. Without a clear-headed leader, both visionary and pragmatic, there was no one to define the plan to carry out the mission—no one to declare the direction in which everything and everyone should be headed, and for what purpose. This vacuum of thought and purpose came angrily into the open in the summer of 1959 at a large party in Dallas. Most of the important space luminaries were in attendance, including von Braun, and a new up-and-comer.

Like von Braun, Christopher Columbus Kraft had been a young boy with dreams. He was only three years old when he suffered a severe burn on his right hand—a burn that would leave a lifelong impediment and scar. Like the Japanese warship that had chosen to mow down PT-109, this event would change the course of a young man's life, and thereby change history. While in college during World War II, Kraft attempted to enlist in the navy. Due to the damage to his hand, he was rejected. Reluctantly, he continued his studies at Virginia Tech. As a result of a wartime urgency for scientists and engineers, the university had been on a twelve-month curriculum schedule, allowing Kraft to graduate in only two years. He was awarded a Bachelor of Science degree in aeronautical engineering.

After graduation, Kraft accepted a job offer from the Langley Research Center, at that time a part of NACA. He excelled at his work, and over the next ten years performed a great many experiments involving rocket planes and wind tunnels. After Sputnik rocked the world, and NASA was created, numerous organizations, including Kraft's employer, were subsumed into the newly formed space organization. From there, fate seemed to play a hand again and again.

Kraft was invited to take part in the new program to launch men into space—Project Mercury. He was soon a member of the Space Task Group and was promoted to be the assistant to Chuck Matthews—the man whose job it was to come up with a manned spaceflight program. Matthews farmed out the task to Kraft, telling him, "Chris, you come up with a basic mission plan. You know, the bottom-line stuff on how we fly a man from a launch pad into space and back again. It would be good if you kept him alive."[2]

Before long, NACA was folded into NASA which, like a bureaucratic black hole, was swallowing and absorbing various departments and organizations as it proceeded to self-create. Kraft had signed on in the early days of rocketry when manned spaceflight was just an idea. NASA, which had existed on paper for only a year, assigned Kraft the responsibility of creating the organization that would perform the actual launching of manned rockets, as well as administer their in-flight plans, programs, and experiments. This organization within NASA would come to be known as Mission Control. Kraft was an excellent choice for the job, even though at only thirty-five years of age he was very young for the responsibility laid on his shoulders. This was Kraft's assignment, and his position on the totem pole, on the night of the Dallas party.

Ever the political schmoozer, von Braun carried his drink and meandered through the crowd toward Kraft, intending to strike up a conversation. He introduced himself, and the two men began speaking. The conversation quickly devolved into a collision course over philosophy. In only a few minutes their discussion elevated to an argument, which then rose to a shouting match. The fuel for this verbal battle was which (or whose) philosophy of manned spaceflight should prevail. In a greater sense it was a fight over the very soul of the program.

Von Braun was a trained pilot who had kept his license and skills current after moving to the United States, and so he possessed a pilot's mind-set, very much like those of the Mercury 7 astronauts. In fact, when all seven Mercury astronauts had an opportunity to meet and talk with von Braun early in the program, they got along famously. The astronauts and the German expatriate saw in each other kindred spirits, and the astronauts left the meeting with a new confidence, not just in the hardware they would fly, but in their master designer.

Such was not the case, however, when Kraft and von Braun met at the party.

Since his youth, von Braun had crafted a vision for what manned spaceflight would be like. In his mind the pilot was everything—a bold and brave individual fully in control of his craft, fully authorized to make all flight decisions, fully responsible for what went wrong, and justly glorified in whatever successes he accomplished. In essence, von Braun envisioned spaceflight as being little different from atmospheric flight—the pilot was in charge. For him the Mercury spacecraft would be a piloted flying machine, like any other. He was shocked to discover Kraft had a completely opposite worldview.

In Kraft's mind, the astronauts would be "employees"—people trained to follow the orders of the land-based mission controllers. Every button they pushed, every switch they flipped, every maneuver they executed, would be done because someone on the ground had planned it out beforehand, then ordered it consummated during flight. For von Braun, this was a serious breach of the pilot's creed—the creed that had endeared him to the astronauts during their visit. Now Kraft was preaching an utterly different gospel, one that would destroy many of von Braun's space travel dreams, and he was angry. Pilots not totally in control of their craft and mission? It was heresy.[3]

In the movie *The Right Stuff*, there is a Wernher von Braun–type character—middle-aged, German accent, scientist's white lab coat—who tangles with the Mercury astronauts over whether the Mercury machine should be a "capsule" controlled from the ground, or a "spacecraft" controlled by its pilot. Most people who saw this movie when it opened in 1983 assumed the actor, Scott Beach (*American Graffiti, Stand By Me*), was portraying Dr. von Braun. But the film credits Beach as playing a character called "Chief Scientist." It's a good thing the filmmakers decided to use a nonspecific character name, as von Braun's ghost would surely have haunted them from the grave for such pilot-unfriendly lines as, "Perhaps there could be a window."[4] The real von Braun would have included large windows in the design before anything else.

As the shouting match between Kraft and von Braun escalated to an ever-higher decibel level, Wernher's wife Maria came to the rescue, gently tugging on her husband's arm and leading him away to another part of the party. But for von Braun, the gauntlet had been thrown. Before the party was even over, he began to plot out in his mind how to blunt Kraft's "idiotic" ideas of total control from the ground.

For his part, Kraft made no secret of how he felt about von Braun. He would later write, "It didn't take him long to tell me that our mission control center concept was all wrong. . . . He had a Teutonic arrogance that he'd honed to a fine edge. He saw himself as the number one expert in the world on rockets and space travel."[5] Of course, at the time von Braun probably *was* the number one expert in rockets and space travel, but to Kraft it was arrogant for an expert to act like one, especially if he disagreed with his plans for Mission Control.

The Dallas party turned out to be a watershed moment for the von Braun camp, which included the engineers and astronauts. It was becoming clear to them that Kraft intended to emphasize the "control" part of Mission Control. Von Braun and the astronauts began an extended philosophical battle with the opposing camp, the NASA brass and managers, over who should control a spacecraft, or if it should even be called a spacecraft at all. In such philosophical battles the winner is usually whoever writes the checks, and it was no different in this case. Von Braun's romanticized ideas of what spaceflight would be—a vision shared by the newly minted astronauts—would have to await some future time. For now, the ground, in the form of Mission Control, would be not only be fully in charge of all launches and flights, but space mission philosophy and policy as well.

Ironically, Chris Kraft had started out with the same philosophy as von Braun and the astronauts. When he first began to pencil out the concept of how to launch a man into space and return him safely to Earth, he assumed that the ground control systems would be similar to those that assisted military planes—a system whose control over those planes and their pilots was minimal. But he soon realized that a program whose flying machines would reach speeds ten to twenty times the speed of sound would often demand skills far beyond those of a mortal human pilot. The rockets would be flying too fast for simple human beings to be able to control all by themselves. It was Kraft who came up with the concept of a ground facility so powerful and overarching that it could run the entire flight without a pilot if it had to. This decision would not endear him to the astronaut corps, but it proved to be the correct one.

Like many aspects of the U.S. space program, the Mission Control facilities had humble beginnings. The first control center was built at the Cape Canaveral Missile Test Annex in Florida and was called the Mercury Control Center, later renamed the Mission Control Center in 1963. It was here that ground support and control of all U.S. manned launches was housed for all Mercury missions, along with the first three Gemini flights. This location had a great many advantages, as all the key engineers and managers involved would be close to the rocket itself for launches. The best minds needed for solving problems as they arose would be on site.

The first control room was a concrete, windowless box designed to protect the men from any explosion or crash. It also protected its occupants from fields of alligators and snakes, hundreds of which lounged and roamed freely on the land upon which the facility was built.

In the early Cape Canaveral days of Mission Control, there were no computers. The first IBM mainframes that would take over the engineering world had only recently come off the drawing board, and they were in great demand. Whoever had the cash, got the computer, and Mission Control was still on a shoestring budget. It would be more than a year before it would snag one for itself. It's difficult to understand today, but in 1958 when engineers in NASA's Mission Control wanted to launch a rocket—one of the most advanced technology endeavors of the day—and they needed to make a mathematical calculation, they used *slide rules*. In its genesis as Mercury Mission Control, each controller was given a workspace that included a primitive small-screen television showing a few white-on-black bits of data, a rotary phone, and an ashtray. Most of the controllers were young men, inexperienced kids right out of college who, like Kraft and Kranz, were forced to make things up as they went along. Manned spaceflight had no manual. The map of the world displayed above and in front of their consoles, designed to show the progress of the space capsules as they circled the globe, included a "toy-like spacecraft model, suspended by wires, [that would be moved by hand] across the map to trace the orbit."[6] Communications with other offices and outposts of NASA traveled over teletype, and as there were no communication satellites yet in orbit, radio communications were strictly line-of-sight.

In November 1960, Senator Kennedy defeated Richard Nixon by a margin of less than 1 percent of the popular vote. The newly elected president cared not at all where NASA's facilities should be located, properly leaving the decisions up to the experts and managers. But after his assassination, the hands-off policy of the White House would change. Vice President Lyndon Johnson was a man who saw great political value in the space program. Having served as both a congressman and senator from Texas, Johnson epitomized the Hollywood image of the tall, burly, gruff Texan man—an image that was well earned in Johnson's case. Enamored of the country's space program since its inception, Johnson had lobbied Kennedy to appoint him chairman of the new National Aeronautics Space Council—a request that Kennedy obliged. Johnson understood the political value of a nonmilitary adventure program run by civilians, and he worked hard to inculcate himself into a program he was certain would have high visibility and positive public awareness. Johnson glommed on to the space program as a pet project and never let go.

As chairman of the NASC, it was Johnson whom Kennedy turned to for advice after the Soviets launched Yuri Gagarin into orbit. The young president was unsure what America's space goals should be, and he asked Johnson and the Space Council to come up with some ideas. What is not well known is that sending a man to the Moon was not Kennedy's idea—it was Johnson's, in concert with the Space Council.[7] After studying what kind of a response the United States should have to the Soviet Union's ever-expanding space program, Johnson recommended swinging for the fence posts—sending a man to the Moon would have a high degree of finality, forever answering the question of which political system was superior.

After President Kennedy was assassinated in November 1963, the U.S. space program had a friend in the Oval Office even more favorable than Kennedy. However, Johnson wanted to make some changes. Like the pure-blood Texas apologist and cheerleader he was proud to be, Johnson soon began to steer assets of the space program to his native state. Though Mission Control could have remained in Florida, the new president wasted no time in using his power and influence to move the command control of all future spaceflights to his home state. In 1965, just in time for the flight of Gemini 4, the new Mission Control Center opened in Houston, Texas.

It was housed inside two rooms of a new building called the Manned Spacecraft Center.[8] It was larger, more modern, and better equipped and designed. And it had computers. All manned space programs and launches, including the future Moon landing, would be controlled from the facility in Houston. The consoles from the original control center in Florida were carted off as museum pieces.

14

SECOND MAN IN SPACE

One good test is worth a thousand expert opinions.

—Wernher von Braun

Once the upgraded Redstone—renamed the Jupiter C—was mated with the new Mercury capsule, NASA decided to hedge its bets by making a few unmanned suborbital test flights first. Unmanned, because in the rocket's maiden voyage a monkey would occupy the pilot seat. One of the early test flight decisions was to fly a spider monkey atop the Jupiter C, followed by a chimpanzee, before they would put a human atop it. It was this decision that caused a great deal of derision between the astronauts and their nonastronaut pilot friends—the presence of an animal sitting in a place meant for a skilled human pilot was a cultural aviation shift of major proportions. Chuck Yaeger took many opportunities to needle his former test pilot compatriots, referring to them as "spam in a can," among other even less complimentary epithets.

But the indignities would only get worse.

James D. "Jig Dog" Ramage was a commander at Naval Air Station Moffett Field near San Jose, California. It was at Moffett that pilots were being trained to transition from piston-powered aircraft to Mach-capable jets. A night fighter squadron called the VF-193, specializing in flying the McDonnell F2H Banshee, was based at Moffett. The squadron was part of Commander Ramage's Carrier Air Group 19.[1] Ramage needed to find pilots with experience flying jets, and there were not many from which to choose. One of them was a young navy pilot whose bravery and exploits were becoming legend. After Ramage arranged for the pilot to be assigned to VF-193, he made the young man his wingman—a great honor.

In May 1954, Ramage and his new wingman were on a routine flight with several trainees at approximately twenty thousand feet altitude when Ramage announced over his radio that his oxygen supply was failing. Without a proper supply of oxygen Ramage soon became disoriented, and any attempt to land in such a condition could have had tragic consequences. The young wingman kept talking to Ramage over their radio, keeping him alert, giving him instructions, and talking him through the landing procedure. With the wingman's assistance Ramage was able to descend to a lower altitude where enough ambient oxygen was available, then safely land his plane.

The name of the young wingman: Alan B. Shepard Jr.

Alan Shepard, like all of the original seven Mercury astronauts, was from a small town. He was born on November 18, 1923, in Derry, New Hampshire. The Shepards were direct descendants of Mayflower passenger Richard Warren.[2] As a boy Alan excelled academically and ended up skipping two grades by the time he reached high school at the Pinkerton Academy. There he developed his love of flight, building a model airplane, and working odd jobs at an airport in exchange for flying lessons. He was only sixteen years old when he graduated in 1940, just after World War II had started in Europe. After a one-year waiting period (owing to his youth) he was accepted into the U.S. Naval Academy. Due to the war's demand for officers, the academy allowed him to graduate one year early as an ensign. During his time at the USNA he would meet Louise Brewer, his future wife.

After the war, Shepard pursued his first love of flying. He transferred to the Naval Air Station in Corpus Christi, Texas, where he was considered so average as a pilot that he was almost dropped from the program. Unwilling to fail, Shepard signed up for additional flight lessons through a civilian pilot school, and eventually was rated good enough to proceed to the Naval Air Station in Pensacola, Florida, for advanced flight training.

The pivotal fork in the road that would determine Shepard's future astronautical career occurred in 1950 when he was selected to attend the navy's Test Pilot School in Patuxent River, Maryland. This fateful move would put him in the position of being eligible to apply for the yet-to-be-created space program. At TPS Patuxent he rose through the ranks, eventually becoming an instructor.

In 1957 Shepard was still in his role as pilot instructor when his name was quietly submitted to NASA as one of 508 test pilots to be considered for its freshman astronaut corps. As the names on the list were whittled down, Shepard's remained, until he was eventually sent an invitation to be one of the original thirty-five candidates to meet at the Pentagon on February 2, 1959. Highly egotistical and brimming with self-confidence, Shepard was not surprised to find himself chosen to be a member of the elite astronaut corp.

From the moment NASA announced the names of its chosen Mercury 7 astronauts, the internal competition had begun. The seven men soon formed a close brotherly bond, but that bond was not strong enough to temper their innate competitive natures. From the beginning, each man lobbied and competed to be the first man in space. The differences in talent and abilities were razor-thin, and choosing one over the other caused a great deal of argument behind closed doors.

Among the astronauts there was a general feeling that John Glenn would be tapped for the first flight, owing to politics entering into many of NASA's decisions. Glenn was very much the way he is portrayed in *The Right Stuff*: a man with wholesome, red-blooded Americana flowing through his veins. Wally Schirra described him as "America personified—baseball, hot dogs, and apple pie."[3]

A small team of NASA executives, led by Bob Gilruth, were in charge of making the flight selections. Cooper, Slayton, Schirra, and Carpenter were devastated when NASA announced the top three astronauts who were in the running for the first three flights: Glenn, Shepard, and Grissom. The men were given the final selection on February 21, 1960, when Gilruth called all seven astronauts into his office and announced that Shepard would fly first, then Grissom, and then Glenn. The astronauts were cordial, each congratulating Shepard and shaking his hand. But secretly they all held some resentment. The reason was simple: it was still anyone's guess as to which country would be the first to launch a human into space—the first-man-in-space record was still up for grabs. Whoever was first in the Mercury rotation stood a chance of being immortal. It would be the kind of achievement that brings forth the type of public accolades, awards, and shiny trophies so prized by test pilots. Only in this case, the trophies might include life-size statues of the man in public parks and government buildings. Whoever was first in space would become a mega-celebrity for many years to come, and each one of the astronauts was well aware of it. So competitive was the internecine rivalry among them that even John Glenn, Mr. Americana himself, was furious at not being chosen for the first flight.[4] *Life* magazine had decided early on to showcase the lives and exploits of the Mercury 7 astronauts, and when the first three flight assignments were announced, the magazine published a story calling Shepard, Grissom, and Glenn the "Gold Team" and Cooper, Slayton, Schirra, and Carpenter the "Red Team." There was no real logic in slicing the Mercury 7 into two such groups—it was purely arbitrary on the publisher's part—just a ploy to sell more magazines. After the article's publication, NASA and the Mercury 7 were forced to do some public relations outreach in order to dispel any misconception in the public's mind that there were two teams instead of just one. In truth, any one of the Mercury 7 astronauts was fully qualified and capable to pilot any of the planned Mercury missions.

Once the astronaut selection and assignment processes were complete, the astronauts managed to gradually put their egos behind them and focus on the upcoming missions. As test pilots, a significant part of the job had been to know their aircraft inside and out. They were, after all, engineers, too. And as they did with airplanes, so too did they with the Redstone, the Atlas rockets, and the McDonnell-built capsule that would carry them aloft. The astronauts threw themselves in with enthusiasm, curiosity, and vigor—educating themselves about the crafts they would be flying. So involved did they become that numerous design changes to the capsule were made based on their recommendations.

As part of their education process the astronauts all assembled at Cape Canaveral on November 21, 1960, to observe the preparation and launch of MR-1 (Mercury-Redstone 1), an unmanned test flight of the hardware configuration that would become the eventual manned Mercury vehicle. The launch was following on the heels of two embarrassing failures. Three months earlier, a Mercury-Atlas rocket exploded mid-flight, then during a November 8 test flight the capsule's escape and jettison rockets fired prematurely sixteen seconds after launch, destroying the capsule. As a result, there was an urgent need and feeling among everyone that the November 21 launch absolutely had to succeed. According to Gene Kranz, "When the Mercury-Atlas 1 exploded in flight, we fell about one year behind in the schedule, so a lot was riding on the first Mercury-Redstone flight."[5]

Yet, success would still elude the fledgling program.

The launch would be a simple affair by later standards—a launch pad, a service tower, and a concrete blockhouse. In addition, three-man radio telemetry teams had been sent out to thirteen listening stations around the world. In the blockhouse, 230 meters from the launch pad, Kranz and the other Americans had to deal with the fact that half the men present were German expatriates, most of them talking in their native language. The countdown was completed without any major incident, and at zero on the clock, the Redstone's main engine fired. Instinctively the video cameraman rotated his camera upward, as if following the rocket, and Kranz and Kraft were riveted to their screens. But there was no rocket in view. The cameraman then lowered his camera back to the pad, and there was the Redstone, still sitting on the pad. Its engine had fired momentarily, then shut down.

As if things couldn't get worse, onboard timers—designed to orderly trigger flight events like a line of falling dominoes—began to activate. Sensing main engine cutoff, the capsule escape tower fired, launching itself four thousand feet into the sky. Then the capsule's drogue chute deployed, followed by the main chute. Offshore breezes caught the chutes, and they billowed open. This threatened to pull the fully loaded rocket over, the result of which would have been a horrendous explosion. In a moment that was almost comical, the capsule then spewed dozens of strips of aluminum foil—designed to help the recovery ship spot it on radar during recovery.

As Kraft angrily tried to find out what had happened, all the replies coming to him were in German.

On a farm about five hundred miles southeast from Moscow, three boys were playing in a field when one of them happened to look up and see a large object falling from the sky. It hit the ground at tremendous speed and its impact created a loud, resounding boom. The object was not far from where they were standing, and they ran over to investigate. Dug part way into the soft farmland soil was a spherical object like nothing they had ever heard of or seen. It had a hatch, which had fallen open, revealing the object's interior. Inside were controls, switches, buttons, and levers. They climbed inside and discovered something else: packets of unopened food. They ripped open the packets and started eating.[6]

Meanwhile, a mere two miles away—unnoticed by the hungry boys—the world's first astronaut, Russia's Yuri Gagarin, floated to Earth on a parachute, having ejected at twenty thousand feet after completing one orbit in the Soviet Union's Vostok 1 rocket.

In the United States there would soon be many more resounding booms: the sounds of doors slamming, books thrown, and objects kicked as news arrived that the Soviet Union had beat America again, notching yet another major space record into the books.

On April 12, 1961, Alan Shepard and the other six astronauts could only watch as Yuri Gagarin made history, becoming not only the first human in space, but the first human to orbit the Earth. Like the American space program, the Soviet program had its genesis as a military endeavor, and like the Mercury 7, Gagarin had been a military test pilot. Upon his return to Earth, the spotlight of celebrity shifted from the Mercury 7 to Gagarin, who rode in a massive congratulatory Moscow parade and received numerous awards, accolades, and medals—one from Nikita Khrushchev. To squeeze every ounce of goodwill out of their space superiority, the Soviet Union would later send Gagarin on a tour of Europe and South America as his country's unofficial ambassador.

The Mercury 7 astronauts had held out hope that one of them would be the first man in space or, even better, the first man in orbit. Now the best they could hope for was just another second-place finish. For men accustomed to being at the top of every totem pole they climbed, it was disheartening. So secret was the Soviet space program that many in NASA and the U.S. government began to wonder if the Soviets already possessed enough off-the-shelf hardware to get men to the Moon. Was the race over but they just didn't know it? With each surprising Soviet space accomplishment, NASA and the American people would second-guess themselves. Everyone wanted an answer: just how far behind in this race are we?

When Kraft reminded Shepard that he had the privilege of naming his craft, Shepard made a decision that was thinly coated with geopolitics. To emphasize the meaning of being the first man in space from a noncommunist country, Shepard dubbed his capsule Freedom 7.

On the eve of NASA's first manned space flight, Alan Shepard, along with his flight backup pilot, John Glenn, went to bed in the spartan quarters of Hanger S, three miles from the launch pad. They were provided two bunk beds. While they slept, the NASA ground crew began filling the propellant tanks of the Redstone rocket. Though it had used liquid oxygen and alcohol in its original version, then later upgraded to oxygen and hydyne for the country's first satellite launch, the man-rated Redstone would use a modified Rocketdyne A-7 engine burning liquid oxygen and the more conventional RP-1, a type of kerosene. RP-1 was safer in formulation, transportation, loading, and use than hydyne, and NASA was taking as few chances as possible.

It was supposed to be simple, short, routine: liftoff, fly a ballistic trajectory for fifteen minutes, splash down in the Atlantic, get recovered by the navy. But like everything in the American space program up to that point, little went as planned. Shepard and Glenn were awoken at 1:00 a.m. and had breakfast of steak and eggs, coffee and orange juice—a meal that would become an astronaut tradition for years to come. Dr. Bill Douglas arrived and gave Shepard his preflight physical. A nurse, Dee O'Hara, attached six biomedical sensors to Shepard's upper body through which Mercury Control could monitor his breathing and heart rate during the flight. Glenn left to go to the launch tower—he had been assigned to perform the last-minute checks of all the spacecraft's switches and meters.[7] Chris Kraft was notified that the weather report was good, and all seemed "go" for a timely launch.

Two hundred eighty miles east of Cape Canaveral, just north of Grand Bahama Island, a U.S. Naval vessel leisurely floated in its fixed preassigned position. The communications room was manned by Petty Officer Stan Burtel, who sat patiently in front of an array of shortwave radio gear, chain-smoking from a carton of Winstons. Like almost every man on the ship, Burtel was an avid smoker, and the tight confines of the radio room contained so much cigarette smoke, it looked like a downsized version of a Southern California summer inversion layer.

The communications coming from Mercury Control at Cape Canaveral were uneventful—just enough chitchat so that everyone knew the communications link was still working. And just in case any interruption did occur, the station had backup radios attuned to backup frequencies. Von Braun's redundancy doctrine extended to every aspect of the space program, including the ships that would be retrieving the astronauts and their spacecraft.

Burtel set his cigarette down as an earnest voice came through the radio's speaker with an important announcement: Astronaut Alan Shepard was on his way to the launch tower.

Dressed in his silver pressure suit, Shepard rode the gantry elevator high above the flat Florida coastline toward the capsule atop the Redstone rocket. John Glenn, as backup pilot, had the privilege of escorting Shepard from their quarters all the way to the capsule. As the elevator came to a stop, they both stepped out and could see the many thousands of cars and spectators lined up at a beach several miles away—tourists who had come to witness history. Across the walkway he could see Glenn standing by the open hatch of Freedom 7, a toothy grin on his face. This was the moment each of them had been training hard for since their selection into the astronaut corps years before. Shepard put one boot in front of the other and approached his craft.

At 5:20 a.m., Gene Kranz received word that Shepard was seated in the spacecraft and would be ready to go on time. Kranz made a note of the time in his log. His expectations and emotions were mixed with trepidation. He would later write, "I felt a shiver. This was history. I hoped that the other controllers were doing a better job of keeping their minds on their work than I was at that instant."[8] With the launch scheduled for 7:20 a.m. there would be plenty of time for Shepard to take his fifteen-minute flight, be recovered at sea, and be washed up and rested long before lunch.

It took more than half an hour to get Shepard fully ensconced and buckled into the capsule, along with a few last-minute checks of the instruments. At 6:10 a.m. Glenn shook hands with Shepard, wishing him, "Happy landings, Commander."[9] The support crew closed the hatch, and Shepard was sealed inside.

Once America's first astronaut was seated and sealed, a pulse of excitement ran through the body of the control room. Kraft had decided that all communication between the astronauts and the controllers would be handled by a single individual. This person became known as the Capsule Communicator, or CapCom for short. Beginning a tradition that would become permanent throughout Mercury, Gemini, and Apollo, a fellow astronaut was chosen to be Shepard's CapCom: Gordon Cooper. Working to keep Shepard's spirits up, Cooper engaged Shepard with some friendly repartee as the countdown progressed. The enthusiasm was soon dampened, however, when a delay in the countdown was announced due to thick cloud cover. As that problem was slowly resolving itself with the Florida sun burning off moisture, another delay was announced due to a computer glitch. When Kranz was informed that the computer problem would take at least ten minutes to resolve, he ordered his team to "take five." Off they went for coffee, donuts, and a bathroom break. When Kraft and crew returned from their break, they discovered Cooper and Dr. Bill Douglas—the flight surgeon—had ordered a comedy nightclub routine be patched into the capsule to help Shepard relax. The dawn of the Space Age had brought with it an opportunistic comedian—Bill Dana—whose standup routine revolved around a reluctant Latino astronaut named José Jimenez. All the Mercury astronauts were big fans of Dana, especially Shepard, and he enjoyed the patch-in immensely. Chris Kraft, however, was not amused.[10]

It was about this time that Alan Shepard came to the realization that there was a major flaw in his pressure suit's design: it was not outfitted with a urine collection system. He desperately needed a bathroom break himself. Shepard had been sitting motionless in the capsule for more than three hours, and the coffee and orange juice had passed through his body. The urge to relieve his bladder was becoming stronger by the minute. He sent word down to Cooper that he needed to exit the capsule for a few minutes to urinate.

It took several minutes for Cooper to receive a reply to Shepard's restroom request. The answer that came down was simple and short: "No." Attempting to put a light touch on the moment, Cooper then mimicked an accent commonly heard around Cape Canaveral—that of Wernher von Braun. "Ze astronaut shall stay in ze nose cone." Shepard and Mercury Control went back and forth over the issue until it was agreed that if Shepard shut off the electronics in his suit (to avoid a short circuit of his biomedical sensors) they would allow him to urinate in his flight suit. And so he did.[11]

This event was followed by yet another delay—the pressure in the liquid oxygen supply lines was elevated beyond flight specs. When informed of the new delay Shepard lashed out, "Why don't you fix your little problem and light this candle!" The problem was resolved, and Chris Kraft began the final GO/NO-GO procedure that would also become a tradition. Each station in the control room was polled one by one for a "GO" or "NO-GO" decision based on what their instruments were showing. One by one each of the controllers responded with "Go, Flight,"[12] and the clock ticked down to zero. At 9:32 a.m., after Shepard had spent more than four hours sitting and waiting, the RP-1 and liquid oxygen valves were opened and the turbo pumps on the Redstone main stage actuated, forcing the propellants into the combustion chamber. There they were immediately mixed and ignited, and the Rocketdyne A-7 engine roared to life. A few moments later the Redstone left the pad and accelerated upward. It pierced a cloud and kept going.

Two significant events followed soon after Shepard's successful flight. First, fellow Mercury 7 astronaut Deke Slayton was grounded from flight status due to a heart defect known as an atrial fibrillation. Second, Grissom's follow-up flight to Shepard's would end in near disaster—after splashing down in the Atlantic the explosive bolts on his capsule hatch prematurely exploded, causing the hatch to sink and Grissom to almost drown. This event would have serious repercussions for the nascent Apollo program.

Though his flight status would be reinstated in time to fly an Apollo mission, Slayton sat out both the Mercury and Gemini programs as a senior administrator in the astronaut office. As a result of his grounding, NASA ended up flying only six Mercury missions instead of the intended seven.

15

KING KONG

There are a thousand things that can happen when you ignite a rocket engine. Only one of them is good.

—Tom Mueller, propulsion engineer

Wernher von Braun liked to doodle and draw whenever a new engineering idea popped into his mind. He would draw what he saw in his head, and what he saw was giant rockets, and the large engines that would lift them into space. And so it was, long before the end of World War II, Wernher could be found in his office late in the evening with pencil and paper, drawing pictures of the kinds of rocket engines he would need to fly him one day to the Moon.

Wernher never got to build those giant engines, but others did—and when they did, he found a way to use them to his advantage. In the mid-1950s, the U.S. Air Force began work on a very large engine—the largest ever conceived or built. Development proceeded steadily, until one day the air force decided to axe the program. The reason? They could not foresee any weapons application for such an engine. And besides, experimentation with clustering smaller engines (perfected by the Russians) was proving more cost effective and reliable. In 1955 the air force decided to end its big-engine program, and everything up to that point went into storage.

Enter NASA. Plans for a lunar-capable rocket started soon after NASA's birth, and right away the organization began fishing around to see if anyone in any other programs or branches of the U.S. government was already working on large rocket systems, especially engines. That's when the old air force mega-engine project was brought to its attention. In 1958 the plans for the air force's engine were taken out of mothballs and happily gifted to NASA, who made sure to get copies to Wernher von Braun. Since his very public success with the Jupiter C and Explorer 1, NASA had learned its lesson: when it came to big liquid fuel rockets, put Wernher in charge. Before long von Braun found himself heading up the design and building of what would become the gargantuan Saturn V.

The contract to build the "big engine" was granted to Rocketdyne in Canoga Park, California, who gave the engine its official name: F-1.

The F-1 was far beyond anything that had ever been successfully built and tested. Its design specs called for an engine that would produce 1.5 million pounds[1] of thrust using liquid oxygen and RP-1 (a blend of kerosene) for a required burn time of 159 seconds. It would burn approximately two tons of propellant every *second*. Each fuel pump would exert the force of thirty locomotives. The combined power of a full contingent of five F-1 engines would generate a power level equivalent to eighty-five Hoover Dams.[2] One could not talk about the F-1 engine without using superlatives.

At least that was the engine that appeared in the blueprints and plans. But in the rocket business, rarely does anything ever go according to plan. At least not right away.

When my rocket-engineering father came home from work each day, he would usually walk through the front door with a cheery, "Hello—I'm home," or some other 1950s television smiley-face greeting. One day—I think I was about nine or ten years old—I remember him arriving in a mood different than normal. He entered the house quietly, stoically. Quite out of character. He went straight to his old wooden desk and sat down, a somber air surrounding him. I probably would have long ago forgotten that moment if not for the conversations that happened next.

He took out his favorite writing medium—a pad of graph paper. He sharpened a pencil, then started writing on the paper. He was utterly lost in thought. When I asked him if he'd like to throw the baseball around (something he always enjoyed doing), he didn't answer. Again, very out of character.

I went to the kitchen to tell my rocket-engineering mother (though at the time, I had no knowledge that she had worked in the rocket business as a chemical engineer). "Something's wrong with dad," I said.

She nodded. "Yes. He's having problems at work."

"What kind of problems?"

"It's an engineering problem—one of those that doesn't seem to have a solution." She dried her hands on a towel and stirred something on the stove she was cooking. "They're having problems with King Kong."[3]

I had seen the original black-and-white *King Kong* movie, so I was familiar with the term. But I was puzzled by what this had to do with my father. I returned to his office and said, "I hear you're having problems with King Kong."

The somber mood vanished, and he laughed. He looked at me and said, "Where did you hear that?"

"From mom."

"Well, she's right."

And that's when I first learned about the F-1 engine—an engine the scientists and engineers at Rocketdyne, my father's employer, had just begun testing on "the Hill"—the Santa Susana Field Laboratory. The SSFL had been built on 2,400 acres of rolling, rocky mounds situated between Simi Valley and our home in Canoga Park, a suburb northwest of Los Angeles. Spread throughout those 2,400 acres were two dozen massive steel girder rocket test stands and a half-dozen blockhouses. Day and night, rocket engines were static tested there, and from where we lived up close to the rocky knolls at the western end of the San Fernando Valley, we could not only hear those engines roar, but feel them as well. At night it was an especially entertaining show as the rocket exhaust flames would light up the sky from just over the ridge. Sometimes we would sit by our pool and just watch and listen.

The SSFL had a culture—a culture of deep engineering and hard science, mixed with a glob of adrenaline and a sprig of sophomoric pyromania. The sandstone hills and knobby knolls of this geologically fascinating land was where the real princes and kings of the nerd world hung out—it was their sanctuary, their palace, their little bit of heaven. But the SSFL had one problem: there was often a great deal of downtime—periods of preparation and setup when technicians were installing rocket engines to be tested, fueling the propellant tanks in preparation for those tests, or fixing a never-ending stream of equipment glitches. As a result, the work environment for engineers at the Santa Susana Field Laboratory was often a lifestyle of hurry-up-and-wait. During these periods of inactivity and boredom, the rocket engineers would pass the time engaging in the most juvenile, frat-boy science nerd behavior imaginable. Bill Webber was one of those engineers. During several of our interviews, he rattled off a number of such supernerd hijinks that he either witnessed or participated in. There was the engineer who loved to play with fluorine. Liquid fluorine is so powerful, it spontaneously reacts with almost everything. This particular engineer would tap out a small amount of fluorine from a tank into a glass tube, then pour the tube's contents onto a slab of concrete, watching gleefully as the fluorine drilled a thin clean hole right through the slab. Then there were the Jeep races. At the SSFL there were a couple of World War II–surplus Jeeps that didn't run very well. Whenever a rocket engine that burned liquid oxygen was about to be tested, teams of engineers would race the Jeeps back and forth in front of the test stands. The liquid oxygen tanks near the stands vented large amounts of pure gaseous O_2, and whenever the Jeeps would drive through the test area, the carburetors would get a massive jolt of oxygen, causing the clunky Jeeps to roar forward like Formula 1 racecars. This was life at the Santa Susana Field Laboratory in the 1960s.

March 6, 1959, would be a great historical marker in the life of the SSFL. It was on this day that the first static firing of an F-1 engine was conducted. The test was orchestrated by a team of experienced engineers and technicians who had seen hundreds of prior test firings. The loud and mighty roars from those engines had become routine to these men—almost ho-hum. But the F-1 was far bigger than anything they had ever mounted on a test stand and so, deciding to take no chances, they intentionally underfilled the propellant tanks to a point where the initial test would run a mere five seconds.

The countdown began, and everyone waited, crossing their fingers. What would happen? Would it explode? Would it function according to plan? When the count hit "zero," a volcano of fire roared from the nozzle as the engine ignited on cue. The thunderous pounding the engineers experienced was far beyond anything they had ever heard or felt. A few seconds later, as the engine cut off and quiet returned to the rocky hills of the SSFL, the engineers stood frozen in open-mouthed astonishment. This was an engine beyond imagination. Someone dubbed the engine "King Kong," and it stuck.[4]

As it turned out, the problems with the F-1 were the same as had been encountered on every large liquid propellant rocket engine since the V-2: combustion instability.[5] Combustion instability is a phenomenon whereby the burning propellants inside the engine cause an oscillation effect. If propellant combustion does not proceed smoothly, pressures inside the chamber alternate between high and low levels at a very rapid speed, causing vibrations and a possible destruction of the engine. Combustion instability is somewhat like the tire on a car that is not properly balanced. Due to the destructive forces created by combustion instability, it is a problem that must be resolved in order for any rocket to be considered man-rated.

Fortunately, this problem had been experienced by engineers often enough in the past that they had a head start on fixing it, though in the case of the F-1 the solution proved elusive, even using tried-and-true methods. Since the scientific theories governing combustion instability were in their infancy, the engineers were forced to use trial-and-error methods, trying various ideas until one happened to work. On the F-1 this turned out to be a set of baffles strategically placed on the injector plate where the propellants enter the chamber. After two years of frustrating tests, Rocketdyne was able to get the baffle arrangements into a just-right Goldilocks formation.[6]

Six years after learning about my father having trouble with the "King Kong" engine, I had the opportunity to see an F-1 test up close and personal. Through Rocketdyne he managed to get clearance to take all the members of my high school rocket club to enter the highly secure Edwards Air Force Base and witness an F-1 test in person. The test stand was in a rocky area with no trees or greenery—just sand and rock. The base commander said we could observe the test by standing on a nearby ridge that overlooked the test stand. My recollection was that we were about two hundred yards away. The test was right on schedule, and soon after we arrived the countdown began.

There are a great many things I don't remember about my teenage years. But I will never forget the day I stood on that rocky ridge and watched the test of that F-1 engine. When the count reached "zero," the bright yellow flame of the burning RP-1/Oxygen propellants roared long and loud from the engine's nozzle. The earth shook, the decibels pounded our eardrums, the exhaust flame was awful and awesome. And yet these words I just wrote utterly fail at a proper description. The live test of an F-1 engine cannot be described in mere words. It cannot be explained using language. No carefully selected group of adverbs and adjectives can do justice. An F-1 engine test is one of those rare moments that are neither seen, heard, nor felt, for anything that can be seen, heard, or felt can be described. An F-1 engine test, on the other hand, can only be *experienced.* I will always be grateful to my father for giving me that experience.

Wernher von Braun had been designing a colossal rocket he called the Nova—a rocket which, had it ever been built, would have been one-third larger than the Saturn V. It would also have required a host of new technologies and development parameters. On the assumption that Rocketdyne would eventually fix the design and performance problems of the F-1, Wernher abandoned the giant Nova rocket concept and replaced it with a more modest proposal—the rocket that would become the Saturn V. Less massive than the Nova, but hefty enough in its own right, von Braun originally dubbed this new rocket the Saturn IV, the "four" referring to the four F-1 engines in its bottom stage. When further calculations revealed that a greater payload and lifting capacity would be required, he simply drew in a fifth engine[7] in the center, and renamed the rocket the Saturn V.

As had been the case since their immigration to the United States, von Braun leaned heavily on the assistance of his German colleagues. In the case of the Saturn V, one of the key architects and players would be a brilliant engineer and close associate from the scientists' days at Peenemünde, Arthur Rudolph. After the war Rudolph had worked alongside von Braun at Fort Bliss, then spent two years working in San Diego for the Solar Aircraft Company. Like von Braun, Rudolph became a naturalized U.S. citizen, and eventually Wernher recruited him back to Alabama. In June 1950, Rudolph found himself working at the Redstone Arsenal in Huntsville as the Redstone missile's technical director. During the 1950s he also performed significant work on the Pershing missile project, for which the U.S. Army would later award him its highest civilian honor, the Decoration for Exceptional Civilian Service.

Rudolph's greatest and most historic contributions would be on the Apollo program. In August 1963, Rudolph was promoted to project director of the Saturn V program, where he not only helped develop the hardware that would get men to the Moon, but also the overall mission plan on how to fly there. Yet unbeknownst to Rudolph, forces were at work behind the scenes, re-examining his Nazi past. It was a past that would come back to haunt him years after the Apollo program had ended.

Once the instability glitches were ironed out, Rocketdyne decided to make a documentary film about the Saturn V and its F-1 engine cluster. It sent out a notice to all its employees, asking if any of them had a backyard pool they were planning on draining soon. As a matter of fact, our family was about to do that very thing, so my father responded to the message. A film crew came out to our house, looked at our pool, and said, "Perfect." When the documentary was released, it showed a man sitting in a small yellow raft floating in our pool. The voiceover says, "The Saturn V gulps a volume of propellant every second equivalent to the amount of water in a typical backyard swimming pool." The water in the pool suddenly disappears, and the man and his yellow raft are sitting at the bottom.

As of this writing, the F-1 remains, more than half a century later, the largest, most powerful liquid fuel rocket engine ever built. Since the F-1's retirement, space launch designs have started to gravitate to a more Russian-like design using smaller but more numerous engines for their bottom stages. The Russian Soyuz has a total of twenty liquid fuel motors running at liftoff. Space X's Falcon Heavy has twenty-seven engines firing in its initial stage.

Today all the remaining F-1 engines have been retired to museums or sold for scrap.[8] Unlike the members of my high school rocket club all those years ago, future generations will probably never have an opportunity to "experience" the firing of an F-1. And in an age of smaller, higher-performing engines using more powerful fuels, such as hydrogen, it is doubtful that anyone will ever again build another "King Kong."

16

CASTOR, POLLUX, AND THE "NEW NINE"

I'm coming back in, and it's the saddest moment of my life.

—Ed White, concluding America's first spacewalk

The "rumored two-man capsule" that Bob Gilruth had been unwilling to discuss with Wernher von Braun was about to go public.

As NASA made its plans for launching mankind to the Moon before the end of the decade, it became obvious early on that an intermediate program between Mercury and Apollo would be needed in order to practice maneuvering and docking strategies required for the Apollo trans-lunar flights. Mercury capsules carried one astronaut, and Apollo would carry three, so it seemed logical that such an in-between project should carry two. This program was dubbed Gemini, which in Latin means "double" or "twin." The name is derived from the third constellation of the Zodiac with its twin stars, Castor and Pollux.[1]

And since NASA expected to launch at least ten Gemini flights, with two astronauts each, there was an obvious need for more recruits. On September 17, 1962, NASA announced its second group of chosen spacefarers. Dubbed the "New Nine," it would be the first time that civilians were among the recruits. And the future of astronaut recruiting could be seen in a new combination of skills—not just test pilot experience, but engineering degrees including, in the case of four recruits, advanced degrees. NASA and its astronaut recruiting program was evolving just as the space program as a whole was evolving. The unmistakable macho, frat-boy image honed so enthusiastically by the Mercury 7 astronauts was intentionally being pushed by NASA toward what it should have been all along: a corps of well-trained, disciplined, highly educated professionals. The Cocoa Beach era of Corvette-driving, street-racing, hard-partying rocketeers was on the wane.

The nine new recruits chosen for the astronaut corps would prove historic in many ways. They were Neil Armstrong, Frank Borman, Charles "Pete" Conrad, Jim Lovell, Thomas P. Stafford, Ed White, John Young, James McDivitt, and Elliot See.

Elliot See was proud to be an astronaut, and excited to be training for his upcoming assignment as command pilot for Gemini 9, now only four months away. It was February 28, 1966, and he and his Gemini copilot, Charles Bassett, were en route to Lambert Field in St. Louis for two weeks of routine simulator training. It was at Lambert Field that McDonnell Aircraft was building the Gemini 9 spacecraft. It was also where the Gemini pilot simulator was located. As was the custom for the astronaut corps when it came to NASA-related travel, the two men were flying "first class": a NASA-owned Northrup T-38 Talon two-place jet fighter. See was at the controls, with Bassett seated right behind.

Also flying a T-38 not far behind was the Gemini 9 backup crew, Tom Stafford and Gene Cernan.

As See and Bassett approached Lambert, See discovered the landing strip was not visible, hidden behind a thick layer of low-lying clouds and fog. Despite the poor conditions, See should have been able to properly land the craft, having logged thousands of hours as a test pilot under many different weather conditions. On the ground, the T-38 seemed to come out of nowhere, its twin jet engines screaming and its afterburners roaring as See, discovering he was too low and not lined up with the runway, attempted to pull up. It was too late for such a maneuver, however, and the T-38 crashed into McDonnell Building 101, killing both astronauts instantly.

In the sky above, Stafford had pulled up for another go around, deeming See's approach too hazardous. When Stafford and Cernan landed a few minutes later they were informed that both astronauts had perished.

An investigation into the crash was immediately launched, with Alan Shepard as its chairman. The board concluded that though bad weather was a contributing factor, the main cause of the crash was pilot error. Stafford and Cernan now became the prime crew for Gemini 9, which brought many other astronauts lower in seniority up a rung or two.

One of them was a new astronaut, chosen in the third recruitment group: Edwin E. "Buzz" Aldrin.[2]

Gemini marked a new era in the Space Race—an era in which Wernher von Braun's predominant influence in so many aspects of rocketry was giving way to the brilliance of a new crop of up-and-comers. As the United States began to catch up with the Soviet Union, so too was the rest of the engineering world catching up to von Braun. Gemini's chief designer would be not a German, but a Canadian, Jim Chamberlain.

Born in British Columbia, James A. Chamberlin, like von Braun, developed a keen interest in flight in his youth. In high school he enjoyed building and flying model airplanes. At the University of Toronto, he obtained a degree in mechanical engineering, then transferred to the Imperial College London for his master's degree. He worked for a time with the British aircraft manufacturer Martin-Baker before moving back to Canada. There he continued to work on British-designed aircraft at Avro Anson. After World War II broke out, he transferred to Clarke Ruse Aircraft in Nova Scotia to work on antisubmarine attack planes. Chamberlin moved around between companies several times before he and twenty-four other Canadian engineers decided to leave Canada and join NASA.

At NASA he rose quickly through the ranks, finally landing the chief engineer job on Project Mercury. He was then put in charge of designing the Gemini spacecraft. Chamberlin hoped to finally make a major leapfrog of the advancing Soviet space program by making the United States the first country to launch more than one astronaut into space in a single spacecraft. Chamberlin and NASA felt Gemini would achieve this. Yet once again the United States would find itself in second place. As NASA was preparing to launch its first of two manned Gemini spacecraft, designed to test launch and re-entry mechanisms, the Soviet Union launched Voskhod 1. It was October 12, 1964, and packed into the Voskhod capsule were not two, but *three* cosmonauts. In its quest to remain the Space Race leader, the Soviet Union was both relentless and successful.

By the time 1965 arrived, the United States and the Soviet Union did have one thing in common: they had each become aware of how complex manned spaceflight could be. Both countries had experienced unexpected failures, and both had learned from hard experience that if there was one endeavor in life that Murphy's Law[3] applied to, it was spaceflight. As Chris Kraft and Gene Kranz worked with the astronaut corps and Mission Control to prepare for NASA's second, and last, unmanned Gemini mission, Murphy's Law kicked mightily into gear.

On the morning of the launch of Gemini 2, everything with the countdown was proceeding by the book. The Titan II rocket—a converted ICBM—lifted off on schedule and without a hitch. Those on the ground in Mission Control, however, weren't so lucky. Despite the mission being unmanned, the launch had attracted a large phalanx of news media. As with previous launches, the media hardware managers had plugged their lights, cameras, and other equipment into the Mission Control electric outlets. Unfortunately, there were so many power-guzzling devices plugged into the system that it overloaded, and the entire main room of Mission Control suffered a massive blackout. Their lights, consoles, computers, and communications systems were all rendered useless. According to Gene Kranz, the room was so dark, "I could not read my stop watches."[4]

Fortunately, the blackout did not occur until right after Gemini left the launch pad, and the launch controllers' responsibilities were lessened now that the new control center in Houston was up and running. Still, for a failure-adverse individual like Gene Kranz this was unacceptable, and he immediately put in a new set of regulations for visiting media. Among the new regulations, all TV, radio, and print reporters who needed electrical power would have to bring their own. From that day forward, portable generators would become ubiquitous outside Mission Control.

The main goals for Gemini were:

1. Practice EVAs (extravehicular activities) or "spacewalks."
2. Practice rendezvousing with a second craft.
3. Practice separation and docking maneuvers.
4. Carry and utilize onboard computer systems for the first time.
5. Test and use fuel cells as an electrical power source for the long duration flights to come.
6. Prove that humans could not just endure lengthy spaceflights, but work and function in them as well.

After the successful launches of Gemini 1 and 2, the first manned Gemini launch was scheduled for March 23, 1965. As the final moments of the count-down approached, the excitement of finally bringing forward the von Braun/Korolev vision of multiperson space travel was palpable. In his memoir, *Failure Is Not an Option*, Gene Kranz writes, "Seated next to Kraft, I was about to enter a new age, not unlike the leap from the Wright Brothers' Flyer to the fighter aircraft of the 1930s, bypassing two decades of normal development. With Gemini we were stepping directly into the future."[5] His comment, "bypassing two decades of normal development," reiterated the seemingly impossible goal set in 1961 by President Kennedy. Great leaps in technology in highly condensed periods of time were needed to land on the Moon within that nine-year time period, and so far NASA had dodged the bullet—they were right on schedule.

The astronauts chosen for Gemini 3 were Gus Grissom as pilot, and John Young as copilot. Grissom had been involved with the design of the Gemini capsule, so it was only natural for him to be assigned to its maiden flight. John Young, plucked from the second batch of astronaut recruits, had been a navy pilot before throwing his hat into the NASA astronaut ring. Pairing up astronauts was a matter of one part science, one part art, and one part psychology. Though there were other men more trained and experienced than he, Young had the skills and personality that seemed to gel well with Grissom. When working together the two men got along well, sharing a keen sense of teamwork, and so Deke Slayton, the man in charge of crew selection for most of NASA's flights, decided to pair them up for Gemini's maiden voyage. According to the Kranz memoir, "The astronaut combination [of Grissom and Young] proved to be a splendid one. Both had only the single desire to fly and were pure joy to work with."[6]

It was a well-established tradition in aviation that pilots held the privilege of naming their aircraft. With virtually all its astronauts coming from the ranks of pilots, it was only natural that this tradition be carried into the space vehicle fleet. With a wink and a nod to the sinking of Grissom's Mercury capsule, Grissom and Young dubbed the Gemini 3 capsule the *Molly Brown*.[7]

The countdown reached "zero" with nary a hitch or delay, and soon Gemini 3 was in the first of its three planned orbits. The mission goals for Gemini 3 were humble: send two men into orbit, have them circle the Earth a few times, prove the Gemini design was spaceworthy, then bring the crew back. By the time they splashed down in the ocean, Grissom and Young had executed a textbook flight.

17

PLAN X

The tight-knit flight controller community started to suspect something was up.

—Gene Kranz

Most of the early astronaut recruits came from the navy, but Ed White was one that had transferred in from the air force. Like virtually all his astronaut brethren, he too had been a test pilot. Born in San Antonio, Texas, he attended the military academy at West Point, after which he was given the commission of second lieutenant. Following flight school, White spent almost four years in West Germany flying F-86 and F-100 Sabres. After returning to the United States, he earned a master's degree in aeronautical engineering and ended up as a test pilot at California's Edwards Air Force Base. When NASA sent out a call for another round of astronaut recruits, White signed on, and ended up being chosen as a member of the "Second Nine." He was subsequently chosen to be the pilot for the second manned Gemini mission, Gemini 4. His copilot would be James McDivitt—like White, an air force recruit.

In January 1965, Ed White showed up for what he assumed would be another day of routine training for his upcoming flight on Gemini 4, then scheduled for late May. Soon after he arrived, Chris Kraft took him aside and said he was about to give White some information that he had to keep secret, even from the other astronauts. The Soviets, said Kraft, owned every single manned spaceflight record in the book. The time had come to put the United States in that record book. White was then informed that during Gemini 4 he would perform the world's first EVA (extravehicular activity)—he would leave the capsule and float in space, protected by nothing but his pressure suit. Kraft asked the new recruit if he would be willing, and White immediately agreed. Kraft then laid out the program he had developed to have White train for the EVA in total secrecy—only a handful of engineers and technicians would know about it.

By mid-March 1965, Ed White had spent more than two months quietly training for the world's first spacewalk. He took the responsibility seriously and had been going at it like a relentless machine. On the morning of March 19, he pulled his car into the parking lot of the Houston training facility. Approaching the entrance, a security guard asked White if he had read the morning paper or heard the news. When White shook his head, the guard handed him the front section of the Houston Chronicle. The headline was as disappointing as it was shocking—the previous day, a Soviet cosmonaut had successfully completed the world's first EVA in orbit.

White silently handed the paper back and entered the building to continue his training.

Despite the loss of what could have been the first major space record for the United States, internal NASA changes would mean a number of firsts on the ground during the Gemini 4 mission. It would be the first mission to fully utilize the new Houston Mission Control complex, the first to use Kraft's twenty-four-hour, three-shift system, and the first in which Gene Kranz would work as a full-fledged, eight-hour-shift flight director. He would do so as the head of White Team.

There were three flight teams set up in the Houston MC complex: Red Team, White Team, and Blue Team. Red Team, headed by Kraft, would cover the eight hours during which the astronaut crews performed their in-flight experiments and flight maneuvers. Blue Team, headed by John Hodge, was the planning shift—they would be responsible for helping the astronauts and the Mission Control crew prepare for work, experiments, and flight maneuvers yet to come. In between the Red and Blue Team would be Kranz's White Team. The White Team would be in charge of getting the astronauts to sleep (a difficult job considering they didn't want to rest, let alone sleep). Then, during the sleep phase, White Team would do a series of checks on the spacecraft to make certain all its systems were functioning properly. It was no coincidence that the team colors were those of the U.S. flag. Likewise, the shift order was no coincidence: red, white, and blue. Within two years, the three teams would be extended to eleven, each with a different color.[1]

The Gemini 4 mission was the first in which flight controllers were no longer referred to by their names, but by their title. Henceforth the astronauts and ground crew would address them simply as "Flight."

A week after the Gemini 3 launch, Chris Kraft brought Kranz into his office. Closing the door for privacy, Kraft let him in on the secret of the planned EVA during Gemini 4. Angry that the Russians had stolen his thunder with their own EVA only two weeks prior, Kraft was determined to put the United States in the record books, come hell or high water. The new plan would be not to perform the world's first EVA, but to perform the longest duration EVA. White would perform an EVA longer than cosmonaut Alexei Leonov's, even if by only one minute. Kraft was tired of the Russians holding all the records—it was time for the Americans to own one, even if it was something so minor. He then informed Kranz that Ed White had already spent the past two months secretly training for the EVA mission. "I want you to write the rules and put together the data package we will need to carry out the mission," said Kraft. "This is risky, but I think it's worth the shot at getting a spacewalk on McDivitt's mission."[2] He again stressed to Kranz his desire to keep the mission change a secret from almost everyone inside NASA. If the EVA had to be scrubbed for lack of readiness, Kraft wanted to scrub it without having it devolve into a public relations mess.

Kranz left the meeting determined to make certain the crew of Gemini 4 would be more than ready. As he began to craft his plan on how to complete that task while keeping the whole thing secret, he decided to label the EVA mission "Plan X." He liked the way the name gave it a secret agent/spy vibe.

As Kranz's wife Marta went about doing all the work to move their home to Dickenson, Texas (about ten minutes from Houston MC), Gene began working two shifts. At 5:00 p.m. he would leave for home and have dinner. Then he would return to the office to work with the Plan X task force. This schedule lasted a full month before others began to suspect something was up. On May 10 a series of routine mission briefings were completed, during which the planned EVA was still not mentioned.

After the May 10 meeting, Kranz called a meeting of the remote CapComs—the astronauts who manned the remote radio outposts in Australia and Hawaii. He gave each of them a double-sealed envelope and instructed them not to open the envelopes until he gave the word at some future date. Sans such instructions, the envelopes would remain forever sealed. Inside the envelopes Kranz had not only placed information about the secret EVA attempt by White, but also another mission addition: an attempt to rendezvous the Gemini capsule with its detached booster rocket. Such a maneuver would not only be good practice for the docking maneuvers required for the Apollo missions, it also would be a "first" that the United States could write into the space record books.

At this time, an inventive sartorial choice would make it into NASA history and culture. In keeping with a desire for team identity, Marta volunteered to sew for her husband a vest to wear whenever he was operating as flight controller. The vest was stark white, matching Gene's team color. He wore it on Gemini 4's launch day, June 3, 1965, working for the first time in NASA's new Mission Control Center in Houston, Texas.

From then on, when working as a controller, Kranz would never be seen without it.

Hours later, Ed White accomplished the no-longer-secret EVA, setting an elapsed time record in the process. However, he was having so much fun that he was reluctant to end it. After the allotted time for the spacewalk had expired, NASA was forced to order him back into the capsule. As he prepared to follow orders, White said, "I'm coming back in, and it's the saddest moment of my life."

That success was tempered by McDivitt's failure to complete an accurate rendezvous with the second stage of the Titan II rocket. NASA was not yet familiar with orbital mechanics, but Gemini 4 yielded some insights that would pay off later with Gemini 6.

For Gemini 5, a Mercury 7 astronaut would once again be paired up with a recruit from the New Nine: flyboy hotshots Gordon Cooper and Pete Conrad. Cooper was a logical choice, as Gemini 5 was tasked with an endurance goal—his Mercury flight logged more orbits and space time than all the other Mercury missions combined. NASA needed to know how long men could endure spaceflight. At that time the Russians held the space endurance record of five days. But for NASA, the magic number was eight—the number of days required to complete a basic lunar landing mission. Was it even possible for humans to survive that long in a weightless environment? Many medical professionals had their doubts. The Gemini 5 crew was tasked with answering the question, so they were assigned (and successfully completed) a record-breaking eight-day mission.[3]

Once the spaceworthiness of the Gemini spacecraft had been proven, along with the ability of humans to live and work in space for extended periods, it was time to get down to the nuts and bolts of maneuverability, rendezvous, and docking—skill sets without which going to the Moon would be impossible.

18

CASTOR, POLLUX, AND THE "THIRD FOURTEEN"

The probability of success is difficult to estimate, but if we never search, the chance of success is zero.

—Giuseppe Cocconi and Philip Morrison

In the mountains behind the enlisted men's housing at Edwards Air Force Base was an unofficial running trail for the military jocks who wanted to stay in shape. It was steep, rocky, and infested with rattlesnakes. In the afternoons it was common for the temperature to exceed 100° Fahrenheit. On just such a day Michael Collins found himself standing on the trail, bent over, hands on knees, and trying hard not to retch. Blessed with a naturally athletic body, he had been able to spend most of his life avoiding exercise. He had also spent a large part of it smoking two packs of cigarettes a day.

As a member of the "Third Fourteen"—the third group of astronaut recruits that followed the original Mercury 7 and the "New Nine"—Collins had gained a great deal of bravado and self-confidence. He had beaten out hundreds of other very qualified candidates. Wasn't he now a member of the chosen elite? It was just a matter of time before he would be tapped for a major space flight assignment—no need to worry.

But he had been keeping an eye on the other members of the astronaut corps. Most of them were leading very different lifestyles and leapfrogging his accomplishments. In an age when smoking was very common, many of the astronauts did not smoke, and quite a few engaged in regular exercise regimens. Frank Borman, Jim Lovell, and Ed White had all been chosen in the second group of astronauts, and it did not escape Michael's attention that men he knew well, and with whom he had come up through the military ranks as coequals, had been chosen ahead of him. Borman and Lovell seemed destined for astronaut greatness, and they were both nonsmoking runners. Collins had met Borman at test pilot school, where they had shared an adjoining desk. And he had known Ed White since their days at West Point, where they met as freshmen plebes. Like Borman and Lovell, White too was an avid runner. Collins had started out at about the same point in life as many of the other astronauts, yet when it had come time to choose the crew for the historic fourteen-day Gemini 7 flight, Collins had to be satisfied with being Lovell's backup.

He tried to hide it, but it stuck in his craw.

One Sunday in the spring of 1962, Michael Collins decided it was time for a change. He woke up that morning "badly hung over with a throbbing head and a throat like an old flue, drier and dirtier than the Mojave Desert sand."[1] He considered his lot in life and his personal health. And he considered his pecking order in the astronaut corps. Michael Collins decided it was time to claim his rightful place in that order. Holding up a pack of cigarettes, he shouted a vow to his wife, telling her that when that pack was empty, he would be quitting cigarettes forever. Two hours later, he took his final drag.

The next day, Collins was scheduled to be the copilot on a four-hour test flight of a B-52 bomber that had recently received a new set of engines. The tremors and fidgeting associated with nicotine withdrawal hit soon after takeoff. "Like a teething baby, I slobbered and gummed my fingertips, my pencils, the corner of my handkerchief. I blew imaginary smoke rings, I inhaled mightily and exhaled in staccato little puffs. I screwed up everything I touched."[2]

Of all manned U.S. spaceflights thus far, Tom Stafford and Wally Schirra's Gemini 6, and its successor 6A, would be the most crucial and historic. For the first time a manned spacecraft would rendezvous with another spacecraft—an important chit that had to be paid in order to ensure success of the upcoming Moon missions. The plan was to launch an Agena rocket (now named the Agena Target Vehicle) minutes before the Gemini launch, then have the Gemini crew rendezvous and dock with the Agena. On October 25, 1965, Stafford and Schirra rode the launch gantry elevator up to the platform where they would enter the Gemini 6 craft. There they were assisted into their seats by Pad Leader Guenter Wendt.

It was an indication of how much the Germans were still involved in so many aspects of the U.S. space program that Wendt would be the last human face every astronaut would see before their spacecraft hatch was sealed shut. Guenter had been a mechanical engineer during World War II, often assigned to maintaining night-flying Luftwaffe fighters. Unlike members of the von Braun group, however, he remained in Germany after the war, not emigrating to the United States until 1949. It took him six years working as a truck mechanic for McDonnell Aircraft before he was able to obtain his U.S. citizenship and be elevated to more important tasks. He began work as pad leader in 1961 with the launch of a chimpanzee named Ham. Thereafter he worked as pad leader throughout the Mercury missions and would continue to do so for Gemini. The astronauts soon began to regard his smiling face and pleasant demeanor as a good-luck charm. But he was stubborn and firm in his place of authority on the launch pad. Pete Conrad once said of him, "It's easy to get along with Guenter, all you have to do is agree with him."[3]

There would be no good-luck charm on October 25, however. Soon after the Agena rocket launched, its second stage exploded, and Gemini 6's rendezvous target splashed into the Atlantic in a thousand pieces.

Stafford and Schirra were given the bad news, exited their capsule, waved good-bye to Guenter, and returned to the elevator. By the time they reached the ground, discussions were already underway as to how to preserve the rendezvous mission. Within hours of the failure, Walter Burke and John Yardley, two McDonnell Aircraft executives, were in Kraft's office with a proposal. Gemini 7 would be ready to launch soon—why not launch both craft at the same time, and have 6 rendezvous with 7? The two craft would not be able to dock, but at least the pilots could practice their rendezvous skills. Schirra and Stafford immediately gave the idea the thumbs-up, but the plan would still have to get the approval of the Gemini 7 crew, Frank Borman and Jim Lovell.

Never before had any country attempted to launch two manned rockets in such close chronological proximity. With the Atlas, Agena, and Titan rockets, the United States had now developed the lifting capacity to catch up with the Soviet Union. But if the risky and ambitious Gemini 6/7 mission could be pulled off, the United States would no longer be playing catch-up. Though the Russians would still manage a few more firsts here and there, the double launch of two manned U.S. spacecraft, which would later rendezvous with each other, would arguably put the Americans in the lead for space supremacy.

To make the post-Agena explosion plan work, numerous changes needed to be made. For one, the Gemini 7 booster was too large and heavy for the usual Gemini launch pad—Pad 6—so the entire operation for both launches had to be moved to Pad 19.

To increase the odds of mission success, Gemini 7 was fitted with a radar beacon to assist in ramping up the precision of the rendezvous. Few of Gemini 7's original mission goals had to be altered to fit the new plan. Gemini 7 had always been planned as another endurance-testing flight—this one for fourteen days. Its orbital flight plan, however, did have to be changed—from elliptical to circular—in order to make the rendezvous easier.

The mission would succeed beyond expectations. After Gemini 7 had been in orbit for eleven days, Stafford and Schirra in Gemini 6 (now renamed Gemini 6A) caught up to and rendezvoused with Borman and Lovell. Schirra piloted 6A, maneuvering for over five hours with Gemini 7, involving distances between one hundred and three hundred feet. What made this world's first outer space rendezvous so unique was the craft's ability to "translate," that is, maneuver without turning, yawing, or banking. The ship could be maneuvered with thrusters in many different positions and axes without the requirement of banking, as in an atmospheric environment. This was a new type of flying, never before experienced by a human being.

Schirra would later describe the experience thus: "Perhaps exquisite harmony is just beyond our reach . . . but I came closest during the Gemini 6 mission. It was when we succeeded in doing the rendezvous. I was at the controls . . . cavorting about, flying rings around Gemini 7."[4]

If anyone had any doubts that test pilots were the best candidates for becoming astronauts, those doubts were assuaged by the events of Gemini 8. Commanded by Neil Armstrong, and piloted by Dave Scott, the mission would record the scariest, most dangerous moment up to that point in the U.S. space program.

The flight plan called for an Agena rocket upper stage to be placed into Earth orbit, after which Armstrong and Scott would be launched in the Gemini 8 spacecraft, catch up to the Agena, rendezvous with it, and dock. It was one of the many practice maneuvers required in order to prepare for the several docking maneuvers required to go to the Moon. After the crew had settled into orbit, Jim Lovell, manning the CapCom console, radioed a warning: "If you run into trouble, and the Attitude Control System in the Agena goes wild, just send in command 400 to turn it off and take control with the spacecraft."[5]

Catching up to and docking with the Agena proved to be straightforward and routine. But within minutes of the docking, the mission began to go seriously awry. Scott, monitoring the "8-Ball"[6] readout that measured their altitude, noticed it was skewing off center. Several attempts with the thrusters to steady the ship and nudge it back in place failed. Under the assumption that Lovell's warning was playing out, they sent the 400 command to the Agena, but the tumble continued. Nothing the astronauts did seemed to work as their craft continued to spiral even more out of control. If the problem continued, there was the very real possibility that the two spacecraft could break apart at the docking port, which could prove disastrous and deadly. In a communication "dead zone" where they had no contact with any of the ground stations, and therefore no way to ask for assistance, and with their maneuvering fuel running low, it was a perfect storm of problems. The two astronauts agreed—the docking had to be aborted. Whatever was wrong with the Agena, it was endangering their lives. Scott threw the switch to undock from the Agena, and the two craft seemed to separate successfully.

Scott operated his maneuvering thrusters, seeking to regain control of the capsule, but to the surprise of both men the spin problem only got worse. Disconnecting from the heavier Agena had exacerbated the spin. That's when the men realized the problem was not with the Agena, but with their own ship.

At that moment Jim Fucci, an astronaut and CapCom stationed on a navy ship in the middle of the Pacific Ocean, started receiving Gemini 8's signal, and what he heard was this: "We have serious problems here. We're . . . we're tumbling end over end up here. We're disengaged from the Agena."[7] With the capsule tumbling, its antenna was in constant movement, making communication spotty. The problem was not in the textbook, and Fucci was at a loss as to how to advise his colleagues. As the tumble intensified, the g-forces and centrifugal forces were putting the astronauts in danger of passing out—a death sentence.

The rule book called for aborting the mission—a decision so serious that it was never considered an option without Mission Control's input. Minutes passed, and the Hawaii ground station became the next communications link. It took Mission Control less than a minute to make their recommendation: abort the mission and initiate re-entry. The astronauts actuated the RCS engines—a use-only-once action, which righted the craft. Mission Control gave them the coordinates to begin their descent and re-entry. As mission pilot, Neil Armstrong would now own a new record: first astronaut to make an emergency landing of a spacecraft. His cool demeanor under pressure would bode well when it came to choosing the crew for the world's first lunar landing.

Though NASA was certain it had worked out the bugs with the capsule's maneuverability that had plagued Gemini 8, Gemini 9 would also prove a major failure, but for a different reason.

Gemini 9 seemed to be cursed from the beginning. First, Thomas Stafford and Eugene Cernan had to take the place of See and Bassett after their deaths in the T-38 jet fighter crash. Then—refusing to give up on the docking mission with an Agena Target Vehicle—NASA launched another ATV on May 17, 1966, for Gemini 9 to dock with. But the ATV malfunctioned and failed to reach orbit. As a backup, NASA mated an Augmented Target Docking Adaptor (ATDA) with an Atlas SLV-3 rocket, which was launched in June 1, 1966, and did reach orbit. But telemetry sent to the ground indicated that something was wrong with the fairing that shrouded the docking port during launch.

Two days later, on June 3, Stafford and Cernan were launched into orbit to catch up with and dock with the ATV. When Stafford and Cernan approached the ATV and the rocket came within view, they were shocked to discover the ATDA's fairing still attached. No docking could be performed under those circumstances. An attempt to salvage the mission was made by performing various rendezvousing maneuvers around the ATV, but an actual docking would have to wait for another opportunity.

An investigation would later determine that an elementary assembly error on the ground by McDonnell Aircraft technicians had caused the jettison failure of the docking fairing. Through a combination of ego and stupidity, those with the proper knowledge of the ATDA's assembly were kept away from the rocket by McDonnell employees who claimed they needed no help.

Gemini 10 would mark the first mission without Gene Kranz at Mission Control—he had been pulled off MCC duty for training and meetings in preparation for the upcoming Apollo missions.

John Young and Michael Collins took the reins of Gemini 10. Once again, the goal was to rendezvous and dock with an Agena rocket, as well as perform at least one EVA. With the exception of Ed White's initial spacewalk, none of the Gemini EVAs had gone according to plan. The astronauts either became fatigued too soon, were blocked from achieving their goals for some unforeseen glitch, or succumbed to some simple error. As before, an Agena rocket was launched into orbit ahead of the two astronauts, who were able to catch up to it in approximately one hundred minutes.

This time, success in rendezvous and docking was finally in the cards. Not only were Young and Collins able to dock with the Agena, and fly it to a record high altitude, they even managed to fly over and rendezvous with the disabled Agena from the failed Gemini 8 mission, still floating in orbit. Though all three of their EVA attempts had been plagued with problems, Gemini 10's elliptical orbit, with an apogee of 763 kilometers, set a new altitude record for manned space. Slowly, steadily, the Americans were pushing aside the Soviet space records and replacing them with their own.

With Gemini 11, Charles "Pete" Conrad (command pilot) and Richard F. Gordon (pilot) would springboard off the success of Gemini 10 to even further heights—literally. They perfected the rendezvous maneuvers with another Agena, taking only one hour and thirty-four minutes for their craft to catch up to and dock with it. Linked together, the combined spacecraft went farther than any previous mission, blowing the Gemini 10 altitude record out of the water with a peak apogee of 1,369 kilometers. The mission even experimented with artificial gravity—using centrifugal force to create a feeling of gravity. The two ships were intentionally put into a mild spin to test out this concept—long theorized by dozens of sci-fi writers and scientists, and made famous by Stanley Kubrick's *2001: A Space Odyssey*.[8]

The mission would have been perfect were it not for yet another EVA problem. Richard Gordon performed two spacewalks totaling two hours, forty-one minutes, but the first EVA, scheduled for two hours, had to be aborted after thirty minutes due to fatigue.

Two astronauts destined for aerospace immortality crewed Gemini 12: James A. Lovell and Edwin "Buzz" Aldrin. NASA was concerned that the EVA still had not been perfected. In order to get over that hurdle, they turned to former Mercury astronaut Scott Carpenter. Though the full story has never come out, Chris Kraft was angry at Carpenter for his flight performance during his Mercury-Atlas 7 flight, a flight which landed 250 miles past its target.[9] Kraft blamed Carpenter for this and other ills of the mission, and permanently grounded Carpenter from further spaceflights. Carpenter left NASA and went on to study undersea science, but he eventually ended up back at NASA, training astronauts how to handle weightless EVAs by training them underwater. Carpenter's contribution would prove fruitful.

Owing to all the errors, successes, and wide experience of the combined Gemini missions, Gemini 12 would be a feather in NASA's cap. Lovell and Aldrin performed the now routine rendezvous and docking with an Agena rocket, as well as a laundry list of scientific experiments and several space firsts. But their biggest success was in finally perfecting the EVA. Carpenter's underwater training drills, combined with redesigned exterior restraints and handles on the capsule, resulted in Aldrin achieving all the mission's EVA goals.

The manned Gemini missions lasted from March 23, 1965, with the launch of Gemini 3, to November 15, 1966, with the splashdown of Gemini 12. It had been an ambitious program that pushed the endurance of everyone to their operational limits—a push motivated by the looming "before this decade is out" deadline. Remarkably, during the year and a half of the U.S. Gemini program the Soviet Union did not launch a single manned space mission. The once-great leader of space exploration had taken an unexpected and unexplained hiatus.

Everyone began to wonder what the Russians were up to.

19

EOR, LOR, AND LEM

One of the many things NASA operations . . . had in common with the military was that rest was a scarce commodity.

—Gene Kranz

In the early days of the space program, as hardware design was first being conceptualized, engineers followed a model laid down early on by science fiction writers and movie companies. The world of entertainment envisioned Moon rockets that were tall, silver, sleek, and aerodynamic, with tapered nose cones and wide tail fins. They would descend to the lunar surface as a single vehicle, then later take off with their structure still fully intact. The astronauts inside the rocket would be strapped down to large, comfortable reclining seats as they fought against the forces of several g's of acceleration. The rocket would then return to Earth looking the same as it did when it left the Moon. Such a scenario was seriously considered for a time—so seriously that NASA even gave it a name, referring to it as "direct descent." Wernher von Braun had once exchanged his engineer's cap for a writer's in 1958, publishing a novella entitled *First Men to the Moon*. In the story a giant rocket, carrying several astronauts, launches a smaller rocket toward the Moon, which lands in one piece, then later takes off and returns to Earth in a single stage.[1]

It did not take long for reality to trump the fanciful imaginations of authors and filmmakers, and direct descent as a viable option was discarded. There were two aspects of landing on the Moon that made a direct descent spacecraft impractical, and far too expensive: the Moon's much weaker gravity, and the lack of any atmosphere. Since there was no atmosphere, there was no need at all for a lunar-landing rocket to have a sleek aerodynamic design. A nose cone and fins would be superfluous—even ridiculous. And utilizing the now well-understood benefits of staging, there was no reason not to leave part of the lunar landing ship behind. Anything that was used up and not needed for the return flight could be left on the Moon as dead weight—no reason to carry it home.

Disposing of the direct descent concept of lunar travel was all a violation of the romanticized vision of Moon and planetary landings served up by Jules Verne and Hollywood—a fanciful image that had been around for decades by the time the 1960s arrived. But it was not just authors and filmmakers that promulgated that vision; they were assisted by the ultimate space seer himself, Wernher von Braun. A fully complete ship landing on the Moon and taking off again was what he had envisioned since boyhood.

Some spaceship of some sort would need to land on the Moon. But what would it look like? Or, more important, what *should* it look like? At first the engineers just referred to this future unknown ship as "the lander." Throughout its design phase, the lander went through many conceptual changes. Every pound that landed on the Moon was a pound that had to first be carried aloft from Earth, then later relifted from the Moon, so weight was a determining factor in almost every aspect of its design.

The lander, if successful in its ultimate spacefaring mission, would score a long list of firsts beyond simply landing on the Moon. It would be the first manned spacecraft designed to be flown exclusively in space, the first such craft to land on a body outside of Earth, the first space vehicle to land upright by its own power, the first manned vehicle to transmit radio and video from another heavenly body, and so forth. It would also set a reliability record as the only component of the Apollo lunar missions to never experience a failure.

The size, shape, and overall design of the lander was governed by one other very crucial factor: the resolution of a heated behind-the-scenes debate.

The one thing everyone agreed on was that the future lunar voyage would require leaving Earth orbit, flying to the Moon, entering orbit around the Moon, landing on the Moon, then taking off from the Moon and returning to Earth orbit. Then, as the final step in the journey, the apparatus would need to be able to withstand the heat of re-entry and bring the astronauts home safely. What kind of vehicle could accomplish all these things? The engineers soon realized that the design of such a vehicle would utterly depend on a decision between EOR and LOR.

EOR, Earth Orbital Rendezvous, was a flight plan proposal that involved two ships being launched into Earth orbit—a "mother ship" and a smaller lunar landing ship. They would rendezvous in Earth orbit, refuel at a prebuilt space station, and then leave orbit together for the Moon. Eventually one or both craft would return to Earth. Many engineers, including Wernher von Braun, strongly favored this plan.

LOR, for Lunar Orbital Rendezvous, was very different. Under the LOR plan both the mother ship and a smaller craft attached to it would travel together to the Moon and orbit there before sending the smaller spacecraft on to its landing. The smaller ship would then leave the Moon and rendezvous with the mother ship. After casting off and abandoning the lander in lunar orbit, the astronauts would head home. Many engineers preferred this plan over the other, and soon NASA was divided into two camps, each vociferously arguing its point of view.

EOR had the advantage of not requiring as big a booster rocket because the two craft would be sent aloft separately. Its disadvantage was the need to first build a space station. LOR would require a massive rocket with incredible heavy lifting capacity, but the space station would be unnecessary.

It is interesting to note that the LOR concept had first been proposed as early as 1923 by German rocket scientist Hermann Oberth. Soon after Sputnik, however, the idea began to quietly regerminate within the halls of NASA's Langley Research Center. Yet despite its advantages, the LOR concept "was to remain something of an orphan within the NASA family at every place except Langley for some time to come."[2]

The EOR versus LOR debate became yet another showstopper in a lengthening line of showstoppers. The debate went on for over a year and, like all good government-funded debates, it generated a mammoth amount of paperwork. One engineer involved with the discussion complained, "If all the paperwork NASA generated were piled up, the stack would reach the Moon long before a spacecraft ever did."[3]

In April 1962, von Braun and his assistants attended a briefing at Space Flight Center. The meeting included proponents of both flight plans and, as usual, the debates got heated. One thing, however, was different: the proponents of LOR had done their homework and had brought mathematics on their side. After hearing them articulate their math and engineering reasons for why LOR should win the day, even von Braun began to weaken. An Apollo program director who had once been an EOR proponent was convinced. He stood up and demanded, "I've heard these good things about lunar orbit rendezvous, and I'd like to hear what son of a bitch thinks it isn't the right thing to do." Von Braun stood up and conceded that LOR did have real advantages, but he stopped short of conceding. The meeting reconvened three weeks later and, to everyone's surprise, von Braun declared, "It is the position of Space Flight Center that we support the

March 16, 1926. Robert Goddard poses just prior to launch of his first liquid propellant rocket. *Courtesy of NASA*

Lt. John F. Kennedy (far right) and the crew of PT-109. *Courtesy U.S. Navy*

March 1946. Wernher von Braun and more than one hundred German expatriate rocket scientists at their new desert home at the U.S. Army base at Fort Bliss. They were secretly brought there as part of Project Paperclip. *Courtesy of Heroic Relics*

May 16, 1946. The U.S. Army and German rocket scientists spent several years at the White Sands Proving Grounds in New Mexico, launching V-2 rockets left over from WW II. Note the color scheme change from von Braun's favored black-and-white design. *Courtesy of NASA*

Rocket pioneer Frank Malina poses next to a WAC Corporal sounding rocket at White Sands, New Mexico. On February 24, 1949, Wernher von Braun and the U.S. Army mounted one of these on top of a V-2 and set a new altitude record of 244 miles. *Courtesy of NASA*

September 12, 1962. With future NASA Chairman Lyndon Johnson looking on, President John F. Kennedy gives a speech before a crowd of thirty-five thousand at Rice University and outlines his plan to send men to the Moon "before this decade is out." *Courtesy of NASA*

The collaboration between Walt Disney and Wernher von Braun used television to bring their vision of future space exploration into America's living rooms. *Courtesy of NASA*

Dec. 6, 1957. After the Soviet Union's dramatic launch of Sputnik two months before, all eyes are on America's first satellite attempt, which ends in a spectacular launch pad explosion. *Courtesy of NASA*

January 31, 1958. Less than two months after the Vanguard debacle, a modified U.S. Army Redstone, renamed the Jupiter C, launches America's first satellite, Explorer I, into space on its very first attempt. *Courtesy of NASA*

The first man in space, the Soviet Union's Yuri Gagarin, confers with "Chief Designer" Sergei Korolev in an undated photo. *Courtesy of Smithsonian Institution*

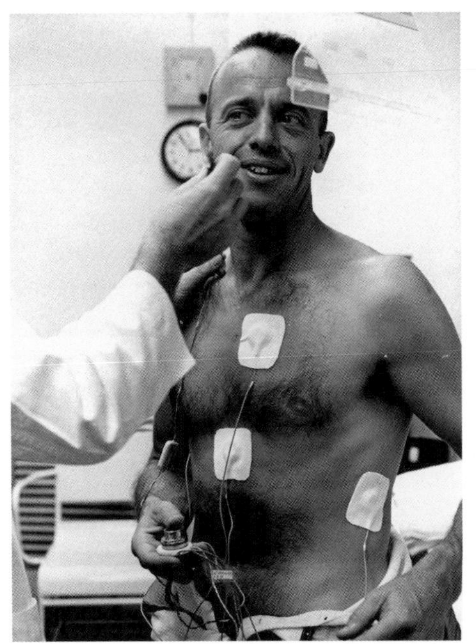

May 5, 1961. Alan Shepard, America's first man in space, undergoes a last-minute physical just prior to his flight. *Courtesy of NASA*

December 15, 1965. A view from Gemini 6 crewed by Wally Schirra and Thomas P. Stafford as they attempt to rendezvous with Gemini 7, crewed by Frank Borman and James Lovell. *Courtesy of NASA*

May 26, 1962. Built by Rocketdyne, the most powerful rocket engine ever built, the F-1, undergoes its first full-thrust, long-duration hot fire test. The Saturn V moon rocket would have five F-1s mounted on its main stage. *Courtesy of NASA*

The Mercury 7 astronauts. Left to right: Wally Schirra, John Glenn, Deke Slayton, Gus Grissom, Alan Shepard, Scott Carpenter, and Gordon Cooper. *Courtesy of NASA*

Wernher von Braun is dwarfed by the F-1 engines mounted on the Saturn V main stage. *Courtesy of NASA*

The Apollo 11 crew. Left to right: Neil A. Armstrong, commander; Michael Collins, command module pilot; Edwin E. "Buzz" Aldrin, Jr., lunar module pilot. *Courtesy of NASA*

July 16, 1969. The Apollo 11 crew aboard a Saturn V lifts off from Launch Pad 39A at the Kennedy Space Center—later renamed Cape Canaveral. *Courtesy of NASA*

July 16, 1969. Wernher von Braun (fourth from left) shares a laugh with Charles W. Matthews, George Mueller, and Samuel C. Phillips shortly after the successful launch of Apollo 11. *Courtesy of NASA*

Michael Collins's view from the Apollo 11 command module on July 20, 1969. The Earth is seen rising over the lunar horizon. *Courtesy of NASA*

July 20, 1969. Buzz Aldrin walks on the Moon beside the lunar module. The photo was taken by Neil Armstrong, who can be seen reflected in Aldrin's visor. *Courtesy of NASA*

Feb. 20, 2012. John Glenn (left) and Neil Armstrong at a fiftieth anniversary celebration of Glenn's historic flight on Friendship 7. *Courtesy of NASA*

lunar orbit rendezvous plan." He later told a close associate that he was in the business of getting to the Moon, not winning arguments.[4]

On July 20, 1962, following lengthy and vociferous battles within the halls of NASA, James E. Webb, NASA administrator, and Dr. Robert C. Seamans, associate administrator, held a press conference where they announced that the LOR concept had won the day. In the end the decision was, like so many other aspects of America's moonshot program, the result of simple pragmatism—there just wasn't enough time and money to build the space station required for EOR.

And NASA had one more acronym for its collection: LEM (Lunar Excursion Module) was shortened to simply LM (Lunar Module).[5]

Once the flight plan was determined, the LM could finally undergo a complete design. Under the final LOR plan the lunar-bound spacecraft would consist of four significant segments, three of which would be discarded at various points during the journey. These four segments included the Apollo capsule that held all three astronauts at Earth takeoff, the service module mounted beneath it, the LM stored inside an upper stage of the booster, and the descent engine package beneath the LM that would eventually be left on the Moon, essentially acting as a minimalist launch pad for the upper manned portion of the LM.

In the end, when it came to actually going to the Moon and landing there, pragmatism outweighed romanticism. It was pragmatism that had enabled LOR to defeat EOR, and it was pragmatism that would define the size and shape of the LM. Bit by bit the idea of the spacecraft's design began to take form, and in the process a number of items were deleted. The large picture windows that appeared in early concept drawings were scrapped as being too heavy and risky (a meteorite as small as a speck of dust could puncture them). And the comfortable reclining seats for the astronauts to ride in: gone. In a move that stunned everyone, NASA decided such seats were unnecessary weight—during the lunar descent the astronauts could simply stand up and man the controls. Von Braun was in shock when he received the news. A spaceship without comfortable, futuristic seats? It was heresy!

The final LM design looked awkward, gangly—like nothing ever imagined by a true space aficionado. When the astronauts of Apollo 9 were allowed to name their craft, they called their lunar module "Spider" for the way its legs splayed out from the body like some giant arachnid.[6] Russell "Rusty" Schweickart, the first trained lunar module pilot, referred to the LM as a "gawky looking bird, but a wonderful flying machine."[7]

Initially two LMs, LM-1 and LM-2, were built just for ground testing, destined never to go into space. LM-1 was put through a myriad of tests; the most anxiety producing of these was the drop test. NASA's specifications required that the LM be able to survive intact and fully functionable after a drop in Earth gravity of six feet. It passed this test admirably—multiple times. The tests on LM-1 proved so successful, in fact, that a decision was made to suspend all planned tests with LM-2 as being unnecessary, and the LM-2 was mothballed. It became what astronaut Rusty Schweickart would later refer to as a "hangar queen"—good for nothing but a dust-collecting museum piece. It is now housed at the Smithsonian's National Air and Space Museum in Washington, DC.

That NASA would put such an expensive piece of hardware on the shelf rather than use it for more testing is testament to how fast-paced the space program was becoming. Every phase of it was accelerating. As many shortcuts that could be taken, without sacrificing a measurable amount of safety, were now being taken on a daily basis. Clearly the engineers were taking President Kennedy's "before this decade is out" dictum seriously, but in retrospect, some of the lessons from the Apollo 1 disaster were already being sidelined.

Eleven companies were given the opportunity to bid on the LM contract, which at the time was expected to cost approximately $350 million.[8] After studying all proposals and bids, NASA chose Grumman Aircraft to be the prime contractor. Grumman was able to snatch the golden ring by being ahead of the curve. A full year before President Kennedy even announced the goal of going to the Moon, a Grumman engineer by the name of Thomas J. Kelly, along with about ten other engineers, foresaw the coming age of manned space travel in a very von Braun–ish way and began to set up a small space division in a company that had never had one. Grumman's bread and butter was jet fighters, but in Kelly's mind the writing was on the wall: there would be money available at some future date for manned spaceflight to the Moon. And so he and his colleagues began laying the groundwork for what hardware might be needed for just such a journey. By the time NASA announced bids for the LM, Grumman's groundwork had already been completed. Their LM proposal to NASA was quick and convincing, and it took their competitors by surprise.[9]

Grumman, however, would not build everything. Rather, the firm became the prime contractor, farming pieces of the design to others. Eventually six other companies would sign on to build parts of the LM: TRW would build the descent engine, Marquardt[10] would build the reaction control system, Bell Aircraft was awarded the ascent engine, and Hamilton Standard would design the environmental control systems. The primary guidance and control were handed over to the Massachusetts Institute of Technology Instrumentation Laboratory, and Raytheon took on the Apollo guidance computer.

LM prophet Thomas Kelly would head up the project at Grumman. Together with several engineers at NASA's Langley Research Center, Kelly had been one of the first proponents of the LOR flight plan, a plan he helped conceptualize and develop. A valedictorian of his graduating class at Ellington C. Mepham High School in Merrick, New York, Kelly would further distinguish himself with a bachelor's degree in mechanical engineering from Cornell, a master's degree from Columbia, and a PhD from Polytechnic Institute of Brooklyn.

As much confidence as NASA had in Kelly and Grumman, they did not allow the Grumman team to have free reign on the design. Instead, NASA insisted on working with Grumman as a partner to create the concept of the world's first space-only spacecraft. The agency would encourage Grumman to come up with some ideas, then it would mold those ideas in its own image based on NASA needs and qualifications. It was a creative partnership in which Grumman created, and NASA re-created. Still, much of the final design came out of the minds of Kelly and his engineering team. Years later in an interview Kelly said, "We just pretty much let our imaginations run wild and let the form follow the function. It just kind of evolved. You basically started out with the two astronauts, and you had to wrap everything around them."[11]

As "form followed function," a peculiar design element of the LM slowly began to creep into focus: a spacecraft without symmetry. Said Kelly, "In the process of simplifying the systems, we realized that we had just fallen into accepting some basic things that weren't necessary, like symmetry. We . . . originally had four-propellant tanks in the ascent stage because it gave us a symmetrical configuration. Then we said, "Gee, it doesn't have to be symmetrical." We could get down to a single tank each for fuel and oxidizer, but then you had to offset them by different amounts,[12] so the LM ended up looking like it had the mumps on one side. So we did things like that, that gradually changed the design."[13] This lack of symmetry, more than any other aspect, would give the LM an outward appearance unlike any craft mankind had ever imagined, let alone built.

The LM was so complex that it ended up requiring the work of over three thousand engineers who amassed an encyclopedic collection of more than forty thousand engineering drawings. One of the well-understood axioms of manned rocket flight was the requirement for double, or even triple, redundancy. Per the "von Braun Doctrine," every critical component needed a backup system to take over in the event of a failure. The drawback to redundancy, however, is that it vastly increases the weight of a spacecraft. Difficult tradeoffs became more and more common as the LM design progressed from idea to concept to engineering to construction to testing. At one point in the LM design phase, things were at a stage where Kelly was ready to lock the blueprints and begin construction. No sooner had he done that, however, then the team discovered the LM would be significantly over NASA's weight specification; it would be too heavy—beyond the capability of the Saturn rocket to lift it into its proper orbit. Kelly initiated a special program he called the Super Weight Improvement Program. After six months of compromising on numerous aspects of the design, the engineers managed to get the LM's weight within required specs.

Bell Aircraft, which had the ascent engine contract, knew that if anything on the LM had to be fail-safe, it was the engine. If it failed to ignite, the astronauts would be stranded on the Moon, without any hope of rescue. They would survive a short while, until their oxygen ran out, and then they would expire from asphyxiation—their bodies forever entombed in the LM. In the interest of reliability, therefore, Bell chose hypergolic propellants for the ascent engine's fuel and oxidizer. Hypergols are propellants that ignite simply by coming in contact with each other, which increases the reliability in one area, but can cause problems in others. For the LM engine Bell decided on dinitrogen tetroxide for the oxidizer, and Aerozine 50 as a fuel. Aerozine 50 is a 50/50 mixture of hydrazine and unsymmetrical dimethylhydrazine, a rocket cocktail originally developed in the 1950s by Aerojet General for the Titan II ICBM. Aerojet had shown that the hypergolic combination was reliable, stable, and could be handled with relative safety. Yet when Bell began testing the ascent engine using these propellants, the motor quickly displayed problems with instability. Testing data revealed the engine had severe pressure spikes during operation. A number of fine engineers worked on the problem for months, without a solution. The problem was so severe that NASA labeled the ascent engine the LM's showstopper—the one item that was not coming along as planned and that was now threatening to derail the entire Apollo launch schedule.

Finally, Bell decided to swallow its pride and bring in some engineers from its biggest competitor: Rocketdyne (a newly named branch-off from North American Aviation). Rocketdyne had experienced combustion instability on many of its engines, including the mighty F-1, destined to be the Apollo booster powerhouse. The firm's engineers had a great deal of experience resolving combustion instability and, once brought in to the project, were able to make a few subtle design changes and get the ascent engine under control. In the end, the engine instability problems and their resolutions gobbled up a full two years' worth of time.

NASA insisted that the LM's systems be tested not just on Earth, but in the cold, weightless vacuum of space, and Apollo 5 would be the shakedown for key components of the LM. The plan called for an eight-hour unmanned mission that would include testing the LM's maneuverability, all controlled internally by the craft's onboard computer. Kranz and his White Team would oversee the launch.

On January 22, 1968, less than two years before Kennedy's decade goal was to expire, Apollo 5 left the launch pad in a picture-perfect launch—not a single hitch. Kranz would later refer to it as "smooth as silk."[14]

Jim Fucci was CapCom that day. When the LM's signal was acquired by Australia, he announced, "Flight, we are Go." Within seconds the LM would ignite its descent engine for its first major test. Just as that was about to happen, Fucci announced, "Shutdown!" The LM computer had detected two errors and had aborted its test. In a few minutes it was determined that some reprogramming would have to be done to the LM computer, but that was just the beginning of the problem. Soon communications between the Mission Control Center and the LM became intermittent, and commands had to be sent up several times before the LM would act on them.

A major engineering juggling act now ensued as MCC decided on new commands, mission objectives, sequence of events, and more. Kranz and his team were able to get all the major goals of the LM in-flight orbital tests successfully completed, but only with ground commands—the automatic onboard system was cut out of the plan. The entire Apollo 5 experience left the ground crew gasping. It all just as easily could have gone south.

The next morning, newspapers across the country announced similar messages: "APOLLO MISSION A SUCCESS, LUNAR PROGRAM ON TRACK."[15]

Thomas Kelly's work on the LM would result in NASA giving him the title, "Father of the Lunar Module."[16] For Gene Kranz, it was just another day of controlled panic.

20

DEATH OF A DREAMER

Things do not happen. Things are made to happen.

—John F. Kennedy

President Kennedy wanted to personally inspect how his "before this decade is out" project was progressing. On September 11, 1962, he decided to visit the Marshall Space Flight Center to take a tour of the facilities and inspect some hardware. There was only one individual with the knowledge and charisma to act as tour guide for such a VIP, and that was Marshall's new director, Wernher von Braun. As the men and their entourages walked through the facilities, von Braun was gleeful in his description of how the president's dream was coming to fruition. Though they had been on opposite sides during the war, Kennedy and von Braun soon formed a close bond.

During the tour, von Braun showed the president a small-scale model of the Saturn rocket that would one day send men to the Moon. "This is the vehicle," he said, "which is designed to fulfill your promise to put a man on the Moon by the end of this decade. By God, we will do it!" The tour put Kennedy in such an upbeat mood that he decided to alter von Braun's schedule, attaching him to his entourage as he moved on to his next stop at Cape Canaveral. With von Braun in the audience, the president spoke to the engineers and technicians, promising them, "We will be first!" Von Braun's enthusiasm was, if anything, quite contagious.[1]

Kennedy returned to Washington and the drudgery of the work of politics, but he could not stay away from the excitement of the space program he had minted. Eight months later, in May 1963, Kennedy was back at the

Marshall Space Flight Center to observe a static test of the Saturn booster stage. Ensconced close to the test stand in a concrete bunker—as close as any human was allowed—the president of the United States stood in awe as a volcano of fire poured from the nozzles of the engines in an ear-pounding, ground-shaking, earthquake-like roar. The immense power of the engines took Kennedy by surprise, and all he could do was gape. When it was over, he shook von Braun's hand and exclaimed, "If I could only show all of this to the people of Congress!"[2]

On the morning of November 22, von Braun was on a NASA plane flying back to Huntsville. He had just finished testifying before Congress about the progress of the country's lunar program. In his pocket was an invitation to a White House reception that he and his wife Maria were looking forward to. President Kennedy wanted to talk some more about the space program, and von Braun was always happy to oblige. During the flight the news came over the radio: the president had been assassinated.

The funeral was held three days later, but Wernher was so distressed, he could not bring himself to attend. In his office in Huntsville, he and his secretary, Bonnie Holmes, watched the funeral on TV. He watched as Jacqueline Kennedy led the funeral procession on foot from Capitol Hill, past the White House, and on to St. Matthew's Cathedral. According to Holmes, von Braun said at one point, "What a waste. What a tragic loss of a friend, and a great leader." Holmes would later write, "that was the only time I ever saw him actually cry."[3]

Wernher felt a need to write to Jackie Kennedy. He wanted to offer his condolences, but he also wanted to give her some good news about how her husband's space goals were progressing. A launch of the first orbit-capable Saturn I rocket was scheduled for two months away. Von Braun waited till after its successful launch, then wrote his letter. Enclosing a photo of the recently launched Saturn I, he finished his letter by letting her know how he felt:

> You have been overwhelmed with condolences from all over the world at the tragic death of your beloved husband. Like for so many, the sad news from Dallas was a terrible personal blow to me. We do not know a better way of honoring the late President than to do our very best to make his dream and determination come true that "America must learn to sail on the new ocean of space, and be in a position second to none."
> With deepest Sympathy, Wernher von Braun.[4]

What was yet to be determined was whether John F. Kennedy's dream of space exploration as a vehicle for world peace would be realized. During his famous September 12, 1962, speech at Rice University he made it clear that the

United States owed the world a duty, not only to be first in space exploration, but to maintain it as a realm of peace.[5] More than a year earlier, at a high-level summit between the United States and the Soviet Union on June 4, 1961, Kennedy had broached the subject of a manned landing on the Moon during a one-on-one meeting with Premier Khrushchev. In that meeting he made the leader of the USSR an incredible offer of cooperation: "Why don't we do it together?" Khrushchev initially agreed, but changed his mind twenty-four hours later.[6]

21

OCEAN OF STORMS

We now knew how to do it.

—William Pickering

Most books on the U.S. space program barely mention the 1966 Surveyor 1 robotic lunar probe mission, if they mention it at all. Neither Michael Neufeld's biography, *Von Braun: Dreamer of Space, Engineer of War*, nor Bob Ward's book, *Dr. Space: The Life of Wernher von Braun*, have a single word to say about Surveyor 1. Neither does Deborah Cadbury in her otherwise comprehensive tome, *Space Race: The Epic Battle between America and the Soviet Union for Dominion of Space*. Gene Kranz's *Failure Is Not an Option* mentions the Surveyor 3 mission, but not the more important Surveyor 1. The probe gets a one-sentence mention in Rod Pyle's *Amazing Stories of the Space Age*. Alan Bean's book on the Apollo programs doesn't cover it, and in fact, I could not find a single book written by any astronaut that includes information on Surveyor 1. Not even a contemporary writer like Neil deGrasse Tyson felt it was important enough to include in his book *Space Chronicles: Facing the Ultimate Frontier*.

Yet without the Surveyor 1 mission, landing men on the Moon would have been extremely risky—a gamble made with the lives of astronauts.

Long before Kennedy's "Before This Decade Is Out" speech, geologists and astronomers argued over the geological makeup of the Moon—what was it made of, and what was the consistency of its surface? Before the president's speech, such conversations were little more than ways to pass time around the water cooler. But once the decision to fly to the Moon and land on it had been made, the lunar soil question required a definitive answer.

And that's when the real arguing began.

Was the lunar surface hard enough to support the weight of a heavy landing craft, or was it so soft the lander would simply sink into oblivion? Or was it something in between? The Ranger impactor probes had answered some of the questions, but not all of them. Before human lives could be risked on a lunar landing, NASA needed to verify the data received from the Ranger probes, and it needed to obtain more detailed information on the quality and firmness of the lunar soil. The "soft landing" Surveyor series of spacecraft were tasked with this mission.

The top priority of NASA at the time was, naturally, the manned lunar landing, and that meant the Surveyor had to be a top priority. NASA soon discovered, however, that it had to overcome a philosophical problem at the Jet Propulsion Laboratory. In November 1964—at a time when Surveyor was little more than a blueprint—JPL had launched Mariner 4, a Mars flyby probe. Telemetry showed that everything on Mariner 4 seemed to be functioning properly, and by Christmas of 1967, JPL fully expected to be receiving some spectacular close-up photos of the Martian surface. At the same time, six more Mariner missions were planned for launch over the next few years. While in Houston, NASA was focused on the Moon, out in Pasadena the agency was focused on the planets. The fact that NASA was funding and launching JPL's probes made little difference—Pickering and his scientists were philosophically looking outward to the solar system. To them, the Moon was already yesterday's news. Adding to the problem was Pickering's propensity to mimic von Braun's outreach mania, spending great amounts of time traveling and making speeches.[1]

Caldwell Johnson, codesigner of the Mercury space capsule, was the engineer tasked with studying how a manned lunar lander should be constructed. He spent many hours stressing over the texture of the Moon's surface. Engineers require data for their designs, and there was almost no data on the quality and structure of the lunar soil. After finding there was no one in the scientific community who could give him a satisfactory answer on the subject, he asked, "How in the hell are we going to design landing gear if the Moon's seas are nothing but pools of dust, and the mountains are nothing but blown-glass fairy castles?"[2]

Back in the USSR, Sergei Korolev and his engineers—still planning on landing on the Moon before the Americans—argued over the same question. One evening, Korolev assembled his best advisors and scientists and demanded they give him an appropriate answer. When they argued and, once again, failed to come up with anything but guesses, Korolev stood up and made an executive decision, declaring the Moon's surface to be solid. As he moved to leave the room, his advisors shouted in protest, demanding to know who would be held responsible if Korolev was wrong. "I will," he said, then took pen and paper and wrote a declaration: THE MOON IS HARD. He then signed it "S. Korolev."[3]

America's space management bureaucracy, however, had no capacity or patience for such executive declarations. The questions regarding the consistency of the lunar surface had to be answered by scientific methods before astronauts could be sent to land there.

In Pasadena, JPL packed up the completed Surveyor 1 probe and shipped it to Cape Canaveral for launching. On board the probe was a series of cameras and a package of scientific experiments. Though the experiments were important, it was the simple cameras that were crucial. Each Surveyor would be equipped with at least one camera pointed toward its landing pads. After landing, the photos of the pads would be transmitted back to Earth by radio telemetry where NASA engineers could get a close look at how far the craft had sunk into the surface. The Apollo program simply could not succeed, or even proceed, without the success of the Surveyor program.

On May 30, 1966—about three years before NASA expected to land its first men on the Moon—Surveyor 1 was launched from Cape Canaveral aboard an Atlas Centaur-D. Three days later it fired its retro rockets and decelerated directly toward the Moon without first being placed into a parking orbit. Ten feet above the surface the descent engine was turned off, and Surveyor 1 was allowed to drop by lunar gravity the remaining distance. NASA had planned four Surveyor missions on the calculated odds that only one in four such missions would be a success. In an impressive display at how far the American space program had matured, the very first Surveyor spacecraft executed a textbook landing.

Even more impressive was its targeting. The plan was to land Surveyor 1 in the middle of a flat area known as the Ocean of Storms. The controllers and programmers hit a bullseye, landing the craft exactly where they intended.

According to William Pickering's biographer, Douglas J. Mudgway, Surveyor 1 "landed gently on the surface of the Moon, activated its cameras and, without fuss or bother, began transmitting the first pictures of the lunar soil beneath its feet. It was accomplished so easily that it almost seemed a matter of routine, rather than being a technological feat of the first order. Surveyor had landed upright, its three landing pads penetrating the lunar soil to a depth of an inch or two with all of its systems in perfect working order. For the next month Surveyor 1 responded faultlessly to over 100,000 commands from its Earth-bound controllers [and it] returned more than 11,000 images of the lunar landscape."[4]

Despite this rosy and sanguine analysis, two of the next three Surveyor spacecraft would crash, pulverizing themselves onto the lunar surface. Space travel was still a long way from becoming "routine."

22

SHOWSTOPPER NO. 2

Rockets are large, rockets are small. If you get a good one, give us a call.

—Daniel G. Mazur, Manager, Vanguard Operations Group

In 1960, Rocketdyne received a contract from the Marshall Space Flight Center in Huntsville to build the world's first large-scale liquid oxygen/hydrogen rocket engine. Von Braun and his team knew that the Moon rocket design then forming in the minds of NASA scientists, as well as their own, would require a high-performance second stage, and they estimated that an oxygen/hydrogen system could offer that performance. Soon after tasking Rocketdyne with the engine, which came to be known as the J-2, Marshall began looking about for a contractor to build the entire second stage. Compared to the few hydrogen rocket systems that had been developed thus far, this rocket would be enormous and, without a doubt, require the creation of a host of new technologies. On the day a meeting was held to release details on the size and scope of that project, thirty aerospace companies showed up hungry for the work. Of those thirty, only seven would eventually submit bids. And of those seven, three would be immediately eliminated from the competition for various reasons. In the end, Marshall awarded the contract to North American Aviation (NAA). At least part of the reason for this decision was that Rocketdyne was a spun-off division of North American, so it was assumed there would be better communication and cooperation between the J-2 engine builders and the overall system contractor. Though this assumption turned out to be correct, other problems began to arise inside NAA with the S-II, as the second stage came to be known.

Initially designed with a cluster of three engines, NASA engineers decided to increase the S-II's lifting capacity by bulking up its size and adding a fourth J-2 engine. Then, mere weeks later, they expanded the design yet again, adding a fifth J-2 to the engine cluster. With each J-2 developing two hundred thousand pounds of thrust, the finalized design of the second stage would enjoy a nicely rounded one million pounds of thrust. These, however, represented only a very small number of design changes made by NASA. North American Aviation would experience many problems with the construction of the S-II second stage—some of its own making—but others would be the result of NASA indecision. The space agency kept changing its requirements for the project—its size, weight, thrust, and more. This then required von Braun's design group at the Marshall Space Flight Center to redraw the blueprints for NAA, which meant the company would have to practically start all over. So many design changes were thrust upon NAA that a running joke began among the engineers arriving to work each morning as they inquired, "So, which S-II are we building today?"[1]

NASA had not been around long, but it had been around long enough to develop a particular culture. One of the facets of that culture was its relationship with contractors. Early in its infancy, NASA had developed a practice of cooperation with its contractors—a relationship whereby NASA experts would oversee and assist its contractors in all projects the agency financed. North American Aviation, however, had been around much longer than the space agency and had developed a culture of its own—one that expected a hands-off approach by its customers. This was the way things had been done with its many military contracts during World War II, and it had worked so well, NAA saw no reason to change. But NASA was a civilian operation and conducted its business by a different set of rules. The result was a clash of cultures and ideals between NASA and NAA that resulted in complaints, arguments, and even threats. At the apex of these battles stood Wernher von Braun, who often found himself at loggerheads with NAA's president, Harrison "Stormy" Storms.

By January 1964, Wernher realized he had another showstopper on his hands. North American Aviation was not only having fabrication problems with the S-II second stage, but was committing the unpardonable sin of not allowing von Braun and his engineers to step in and assist with advice. Over the next six months, the working relationship between von Braun and Storms had deteriorated to the point that von Braun was inquiring how Storms might be ousted from his position. At a Marshall Space Flight Center board meeting in September, "Dr. von Braun emphasized that the S-II is the most important single problem confronting MSFC, and therefore carries the overriding priority."[2] At one point, the NAA board chairman, Lee Atwood, visited the von Braun team in Huntsville to work out some of the problems. Atwood attended that meeting because Storms was in the hospital due to a stress-induced heart attack.[3]

Von Braun's ire was amplified not just by S-II construction delays, cost overruns, and fabrication flaws—of which there were many—but also by test results he felt were catastrophic. Liquid fuel rocket systems are always tested with water before they are ever tested with volatile propellants. In a standard "water flow test," the propellant tanks are filled with water, then the engine's turbo pumps are turned on, allowing all the water to be pulled out of the tanks, through the pumps, and out the nozzles. The purpose of these tests is to make sure there are no leaks, all the plumbing functions properly, and the entire system can withstand the pressures of operation. During what was supposed to be a routine water flow test of a completed S-II in September 1965, the entire apparatus broke apart, collapsing under the mass of tons of water. To save weight, the rocket had been built to the very edge of structural integrity—or over the edge, as it turned out.

Despite von Braun's lack of confidence in Harrison Storms, the NAA president and his team eventually resolved all the technological challenges, fixed the many construction problems, and purged all the S-II gremlins, turning out a final Saturn second stage product that was nothing less than an engineering work of art.

23

A ROUGH ROAD

The desire for safety stands against every great and noble enterprise.

—Cornelius Tacitus, Roman historian

In Houston, the astronaut office was quiet and almost deserted. It was January 27, 1967, a late afternoon on Friday—the day when Deke Slayton, the manager of the office, and his staff would gather for a weekly briefing. Alan Shepard was responsible for attending on behalf of the astronaut corps, but he was away from the office. The presence of a senior astronaut was required, and Michael Collins was the most senior one present, so he attended in Shepard's place. As he took his seat in the briefing room he readied his notepad and pen, prepared to take notes for what was usually a boring hour or more. Deke entered the room, and the meeting started.

The room had a desk off to one side, and on that desk was a red phone. It was referred to as the "crash phone"—it would only ring if there had been a pilot/astronaut-involved crash or life-threatening accident. All upper-level NASA administrative offices had one. Except for when it was occasionally tested, Deke Slayton's red phone had never rung. As the staff briefing progressed the red phone sat on the desk as it always had: quiet, still, unnoticed.

Forgotten.

Then, a few minutes into the planned agenda, the red phone rang.

Gene Kranz and his wife were getting dressed in preparation for a dinner date—a rare night out without the kids. Suddenly there was an insistent pounding on his front door, which Kranz at first took to be an early, and overly enthusiastic, babysitter.[1] But when he opened the door, standing before him was fellow flight controller Jim Hannigan. As Hannigan stepped through the door his first words were, "Have you heard what happened?" Not waiting for a reply, Hannigan stepped over to the television set and turned it on. "They had a fire on the launch pad. They think the crew is dead."

Kranz had left the Cape for the simple reason there was nothing for him to do. Apollo 1 astronauts Gus Grissom, Ed White, and Roger Chaffee had spent the day performing tests aboard the CSM (Command and Service Module), which was mounted atop the Saturn V.[2] They were to be the maiden crew—the first astronauts to fly Wernher von Braun's dream machine into orbit. The rocket was standing on its launch pad, but since the tests were preliminary, its propellant tanks were empty. They were still two days away from the actual launch and, at that point in the countdown, everything was considered to be "low risk." Everything on their schedule was routine—nothing that required Kranz's presence. At Hannigan's announcement, the first thing Gene Kranz imagined was an explosion of some sort—but minus the rocket's propellants, what would cause such an explosion? What possible explanation could there be for a deadly fire during routine, prefueling tests? There were explosive bolts everywhere holding the stages and superstructure together, and a few other pyrotechnic odds and ends, but nothing he considered seemed likely to cause the deaths of all three astronauts.

As Gene's wife Marta came downstairs, Jim's wife Peggy ran in the front door crying. On the television news channels, only vague, cursory accounts were available—merely that some sort of accident had occurred. Though Gene could not yet be aware of it, Chris Kraft had already put the Cape on information lockdown, including all outgoing phone calls.[3] Gene grabbed his ID badge, jumped into his Plymouth station wagon, and drove at high speed toward the Cape. He drove so hard and fast that he would later write, "I practically dared a cop to get in my way."[4]

In Houston, Deke Slayton set the red receiver back down on its cradle. He turned to his staff and reported, "Fire in the spacecraft." Since everyone present was well aware of the Apollo 1 capsule then undergoing tests at the Cape, there was no need to explain which spacecraft he was referring to. Everyone sat stoically, wondering what it all meant. As the whispering started about possible scenarios and outcomes, the red phone rang again. A minute later he left the meeting without explanation.

Arriving at the Cape, Gene discovered that Kraft had locked things down so tight that security guards would not allow him into the building. He ended up having to bluff a guard at the service elevator behind the building in order to get in.

On the second floor he approached Kraft, Slayton, and Hodge, who were talking to the Cape's surgeon. The first utterance he overheard was, "It was gruesome." At the consoles, flight controllers were already pouring through their telemetry data, attempting to figure out what had happened. Gene was then filled in; an intense fire had swept through the capsule, immolating the astronauts in seconds. According to Slayton, it was so hot that there was "molten metal dribbling down the side of the service module."[5]

The accident left everyone flummoxed. From the early days of the manned space program it had always been assumed that if astronauts lost their lives, it would be during a launch, during an orbit, or during re-entry—not casually sitting on a test stand 320 feet from the ground atop an unloaded rocket.

The official name of the Apollo 1 craft at that time was AS-204, and the test being performed was referred to by the engineers as a "Plugs-Out Integrated Test." As the name implied, the test involved unplugging the Saturn V umbilical cord, thus disconnecting the rocket and its capsule from the GSE (ground support equipment). The purpose of this test was to see how the entire craft functioned as an independent machine.[6] A simulated countdown and practice of prelaunch procedures, which began before dawn, was a major part of the test. The crew was not required to be present at first, and so they did not begin the "ingress procedure" into the capsule until about 1:00 p.m.

As the simulated countdown progressed, a few minor glitches were encountered. The crew reported an odd smell inside the cabin, and their communication links with the ground control crew kept going in and out. Hours went by, but generally everything seemed routine. Each astronaut was hooked up to sensors that monitored their heart and breathing rates—even how often and how much they moved their muscles. The telemetry data from these measurements was routine and showed no anomalies.

The first indication that something was wrong occurred at 6:31 p.m. It was a voice of one of the astronauts over the glitchy com link reporting, "Hey—fire." It was followed a few seconds later by, "We've got a fire in the cockpit." Events quickly escalated, and their voices became more frantic. At first the crew reported an attempt to fight the fire, then an attempt to open the hatch and exit the capsule.[7] On their video monitors, the ground crew could see the flames. At first it did not seem like a major conflagration, but within thirty seconds, the fire became intense to the point the flames were all that could be seen. Over a communications link that now, ironically, was working fine, the men could be heard screaming.

Outside the capsule the support crew attempted to open the hatch, but they were ill-prepared, untrained, and not equipped for this type of emergency. Flames and thick smoke were shooting out between the melted seals of the capsule, and its surface was far too hot to touch. The support crew was so poorly equipped for such an event they did not even have oxygen masks handy, which was causing them to retreat due to smoke inhalation. This, together with the flames and heat, prevented them from working quickly to open the hatch. The crew did what they could, but within the capsule the fire spread too fast—in less than two minutes, all three astronauts were dead.

North American Aviation manager Donald Babbitt was a member of the gantry support crew that succeeded in opening the exterior hatch and getting the interior hatch at least part way open. He would later testify to the Apollo Review Board, "My observation at the time of hatch removal was that the flight crew were dead and that the destruction inside the Command Module was considerable." Babbitt continued to help with support operations before he was forced to report to the base medical personnel for treatment for his injuries—smoke inhalation and flash burns. It seemed as if everything that could go wrong was going wrong. The support crew, lacking any sort of fire and smoke suppression equipment, had immediately called for the fire department to be sent up. But when the fire crew arrived at the bottom of the launch gantry, they were forced to take the slow elevator up to the capsule because the fast elevator was out of commission.[8]

By the time Babbitt left, the firemen had reached the scene. One of them, James Burch, took over at the hatch. After getting it fully opened, he peered inside the capsule. The damage and chaos were so extensive, and the bodies incinerated so extensively, that his first report was, "There's no one in there." Upon closer examination he found a boot, pulled on it, and found a leg. Three doctors arrived a few minutes later—also on the slow elevator—and declared the crew to be deceased.[9]

Two years earlier, NASA had learned a sad and difficult lesson about the proper administration of public relations. On the morning of October 31, 1964, Ted Freeman, a new astronaut chosen for the third astronaut group, was returning from a training session in St. Louis and was about to land a T-38 jet at Ellington Air Force Base.[10] Moments before his wheels were to touch down on the landing strip, a Canada goose collided with his canopy, throwing shards of the canopy into the jet's engine. The engine cut out and the jet fell to earth, killing Freeman instantly. Ted Freeman was the "first astronaut to lose his life in the service of the space program."[11] What happened next highlighted weaknesses in the NASA chain of command. Freeman's wife Faith was home at the time, going about her daily routine when she heard a knock on the front door. Upon opening it, she was confronted by a news reporter who asked for her comment on the crash and tragic death of her husband. This was how she found out about the accident. After that incident, NASA vowed to never allow a spouse to hear such news from anyone other than a NASA administrator or, preferably, a fellow astronaut. For this reason, all eyes in the briefing room turned to Collins.

Like Freeman, Alan Bean had been chosen as part of Astronaut Group 3, aka the Third Fourteen. It was Bean that answered the phone when Collins called the astronaut office to begin the process of notifying the families of Grissom, White, and Chaffee. Bean said he would find astronauts to do the notifications, but asked Collins to remain by the red phone so that they would have access to the latest information. But minutes later Bean called Collins back with a problem; he had found astronauts and wives willing to immediately visit and notify the Grissom and White households, but had not been able to find a third astronaut to visit Roger Chaffee's wife, Martha. Mike and his wife Pat lived just three doors down from the Chaffees, so Collins informed Bean that he would go. He then got in his car and "drove very slowly the mile and a half to her house."[12]

When Collins arrived at the Chaffee home, he found Al Bean's wife Sue already there, guarding the telephone and screening calls. Several other neighbor wives were also there, so Martha knew something terrible was up.

"Martha, I'd like to talk to you alone," said Michael.

Like the astronauts themselves, the thought that some sort of disaster could happen during a simple test of a grounded rocket was something their wives never considered. All the dangers, everyone had assumed, involved flight—the launch, the orbit, the landing. Though she knew the news was bad, Martha was unprepared for what Michael now told her.

NASA wasted no time in ordering an official investigation. Before the day was out, the NASA Deputy Administrator ordered the establishment of the Apollo 204 Review Board. The Review Board convened the very next day, January 28, 1967, on the grounds of the Kennedy Space Center. Dr. Floyd L. Thompson, director of the Langley Research Center, was its appointed chairman. One of the Board's first actions was to divide the responsibilities of the investigation into twenty-one Task Panels, each responsible for reporting back to the Board information on various aspects of the disaster.

The Review Board began by overseeing the complete disassembly of the command module. Each part (and there were thousands of them), no matter how large or small, was photographed, studied, and tagged. They were then sealed in plastic for possible future re-inspection and reference. That process took two months. Despite the exhaustive nature of the investigation, utilizing hundreds of experts, the Review Board was unable to make an exact determination of the cause of the fire. In the end, the board was forced to rely on probabilities. The final report concludes that the most probable cause of the fire was "an electrical arc in the sector between -Y and +Z spacecraft axes. The exact location is [likely] near the floor in the lower forward section of the left-hand equipment bay where Environmental Control (ECS) instrumentation power leads into the area between the Environmental Control Unit (ECU) and the oxygen panel." For those who had raised the specter of possible Russian sabotage, the report concluded "No evidence was discovered that suggested sabotage."[13]

The Review Board's final report enumerated a long list of observations. Among them were that NASA had become complacent about the risk of the ground tests, that the tests themselves were inherently hazardous, that neither the ground support crew nor the astronaut corps were trained for such an emergency as a capsule fire, that the capsule was improperly designed from the start, and that the capsule hatch procedures were too cumbersome to allow an emergency exit.

The report's list of recommended design and procedural changes was extensive. One item on the list was creation of a better exit system for all future Apollo capsules—one that would allow either the astronauts or the support crew to open the outer and inner hatches faster to allow the crew a safe and speedy exit. This, of course, would do no good during an actual mission. Once aloft, above the Earth's atmosphere, there would be no place to escape to. The capsule had to be independent, and it had to be, more than anything else, fire safe. Also on the list of recommendations was a suggestion that future missions not require a 100 percent oxygen atmosphere—an obvious fire hazard.

The U.S. Senate held hearings on the disaster, which came to its own findings and conclusions. Unlike the Review Board, the congressional hearings dealt with President Kennedy's original goal of landing a man on the Moon "before this decade is out." Of the president's deadline, the hearings concluded that it was "essential to the planning process . . . to maintain a vigorous and competent organization." In other words, a big delay would cost more money than we're willing to give NASA. However, walking a fine line between budgetary limits and public opinion, the hearings also concluded, "Safety must be considered of paramount importance in the manned space flight program even at the expense of target dates."[14]

As a result of all the hearings, the Review Board, astronaut input, and contractor suggestions, the Apollo command module went through a major top-to-bottom redesign. The entry/exit hatch would be made of aluminum and fiberglass, and would be easy and quick to open. The internal oxygen atmosphere was changed to 60 percent oxygen, 40 percent nitrogen, all at a lower pressure of five pounds per square inch (psi). Dozens of smaller fireproofing changes were made in the cabin, even changing the control switches from plastic to metal. In addition, the support crew that worked on the gantry were given an array of fire prevention and firefighting equipment, as well as extensive training for fire emergencies.

The effort proved fruitful—never again would there be a fire, or even a hint of one, in any Apollo capsule.

The autopsies of the astronauts were eventually made public. Careful examination of the bodies determined that despite extensive skin burns, the most probable cause of death was carbon monoxide poisoning due to smoke inhalation.[15] Essentially, they asphyxiated before the flames had a chance to kill them. Either way, the astronauts never stood a chance of surviving the accident.

That Gus Grissom was on board at the time of the Apollo 1 fire was ironic, though the irony wouldn't be noticed until later. After Grissom's hatch prematurely blew open on his Mercury splashdown, NASA had reengineered its hatches—including those on Apollo—to make them sturdier—that is, harder to open. In hindsight, it's odd no one thought of the consequences—that in an emergency, a more difficult-to-open crew hatch could be deadly. Adding to that irony was Grissom's own prophetic statement a month before the accident: "There's always a possibility that you can have a catastrophic failure, of course; this can happen on any flight; it can happen on the last one as well as the first one. So, you just plan as best you can to take care of all these eventualities, and you get a well-trained crew and you go fly."[16]

The managers, engineers, technicians, and employees of NASA would never be the same. The deaths of their colleagues weighed heavily on all of them as they shared group guilt. These feelings were especially acute in Gene Kranz. He would later write, "I felt that I had personally let down the crew of Apollo 1." Not long after the fire, Kranz called a meeting of his Mission Control staff. He took the stage and spoke. He talked about the "terrible tragedy" and its "devastating setback." But he assured everyone that this would not end the program, or their mission. He made it clear that it was their responsibility to "not let the Apollo 1 crew die in vain." "Spaceflight," he continued, "will never tolerate carelessness, incapacity, or neglect. Somewhere, somehow, we screwed up. It could have been in design, build, or test. Whatever it was, we should have caught it."[17]

The audience was quiet and still during his comments. He let everyone know that things were about to change. "From this day forward, Flight Control will be known by two words: 'Tough and Competent.' 'Tough' means we are forever accountable for what we do or what we fail to do. . . . 'Competent' means we will never take anything for granted. We will never be found short in our knowledge or skills. Mission Control will be perfect." He then ordered everyone to return to their offices and write the words "Tough" and "Competent" on their chalkboards and bulletin boards—and never erase them. "Each day when you enter the room," he declared, "these words will remind you of the price paid by Grissom, White, and Chaffee. These words are the price of admission to the ranks of Mission Control."[18]

In any disaster, finger-pointing and blame assignment is inevitable, and the Apollo 1 fire had plenty of both. There were many theories as to the cause, none of which were proven. In the final analysis, the real culprit was complacency. NASA had utilized a 100 percent pure oxygen atmosphere in every capsule of the Mercury and Gemini program without incident, and so they had made the decision to continue to do so with the Apollo, despite its capsule's greater complexity. And in Apollo, the oxygen danger level had been inadvertently heightened—the cabin pressure had been set at 16 psi (even higher than sea level), thus increasing the fire hazard.

Like Collins, Ed White had been a West Point graduate. Unfortunately, White's funeral was to be held at West Point the same day that Chaffee's was scheduled at Arlington. Collins felt torn; he had been a classmate of White, but he also felt a closeness to Chaffee because they had been hired as astronauts on the same day, and it was he who had borne the duty of breaking the tragic news to Martha. So, on the day of the funerals, Mike Collins was at Arlington.

Two and a half years later, when Apollo 11 landed on the Moon, the astronauts carried with them a diamond-studded pin that originally had been intended to fly on Apollo 1. Today there is a plaque mounted at Cape Canaveral dedicated to Grissom, White, and Chaffee. Its inscription reads, *AD ASTRA PER ASPERA* (A rough road leads to the stars).

Though the deaths of astronauts have always garnered headlines, the injuries and fatalities of lesser-known space program workers—people the media paid no attention to—were lengthening in number and severity as well. The list of such incidents is long, but here are two examples.

One day a worker was on one of the Cape Canaveral test stands when he noticed what looked like a small icicle forming on one of the steel flooring panels. Though it may have looked like water ice, it was actually a buildup of liquid oxygen leaking from a pipe high above. The man nonchalantly kicked the "icicle," thinking he was merely removing a small hazard. Upon kicking it a small spot of oil on his boot came in contact with the oxygen. The result was instantaneous—a huge chemical reaction that blew the man's leg completely off.

In another incident, workers were doing maintenance work inside of one of the Saturn V's propellant tanks. The workers were required to shimmy through a narrow opening, get fully inside the tank, and thoroughly clean it. Somehow a worker entered one of the tanks when it was filled with gaseous nitrogen, and the man quickly asphyxiated.

A "rough road," indeed.

24

REBUILDING

At the time of the accident, every element of the program was in trouble.

—Gene Kranz

With the Apollo 1 disaster, NASA lost three crucial elements in its goal to land on the Moon by the end of the decade: its best three-man crew, a fully constructed Apollo capsule, and the trust of the American people. All would have to be reassembled and rebuilt. NASA leaders were politically astute enough to realize the deaths of Grissom, White, and Chaffee had the capacity to cause a huge delay in their program goals. There would be investigations and incriminations and Senate panels and so forth—all designed to uncover facts while slowing things down. Chris Kraft and Gene Kranz, however, were determined to stay on target to the greatest extent possible. While they mourned the loss of their colleagues on one hand, they quietly pushed their contractors to stay on schedule on the other. The only exception to this, of course, was the redesign of the Apollo capsule. Since none of the other Apollo capsules had yet been completed, it was not too late to make the crucial mid-course design and manufacturing corrections that would be required. NASA administrators therefore had two missions—to keep the construction of all rocket and mission hardware on schedule, and to fix the Apollo capsule's design flaws as quickly as possible.

To accomplish this, several planned missions would be permanently scratched off the list, and Apollos 2 and 3 were the first to go.

During the Gemini phase of the lunar landing program, it had been determined that duplicate missions were a luxury NASA could little afford. When

Gus Grissom was sent into space on a suborbital flight on July 21, 1961, it was essentially a duplication of the Alan Shepard flight—nothing new had been learned (other than that the Mercury capsule escape hatch design was possibly flawed). From then on NASA began to form an unofficial policy of no more duplications—every time a man went into space, some new aspect of long-range lunar-centric planning would be tested. Even so, the Apollo 2 mission had been slated to be a duplicate of Apollo 1. At some point prior to the Apollo 1 disaster, Dr. George Mueller, associate administrator of the Office of Manned Space Flight, decided to axe the Apollo 2 mission on the grounds of redundancy. Now, with the loss of Apollo 1, there was no vehicle available for the intended tests of those missions.[1] NASA had started its Apollo program planning with two baby-step missions, and ended with neither.

Ironically, this only solidified into NASA policy the concept that duplicate missions could no longer be afforded, whether due to time or budget restrictions. When the Apollo 1 disaster threatened to blow up the entire lunar landing schedule, small incremental progress was taken off the table. There would be no more baby steps of improvement—henceforth all manned flights would seek some major leap in testing and discovery. The clock was ticking, the end of the decade was approaching—time was the silent enemy. The result was the cancellation of a third Apollo mission—Apollo 3. The next Apollo spacecraft to fly would be named Apollo 4.

Balancing the need to be safe and cautious against the fact that success in the space program required great risk, NASA decided to send the next two Apollo capsules up on Saturn V rockets—both unmanned. These missions were capable of testing some pieces of the mission hardware remotely, and thus with greater safety. Apollo 4 would mark the maiden voyage of von Braun's gargantuan Saturn V rocket system—a rocket so large in power and scope that NASA decided that in the new postfire world, it would be prudent to test the entire system a few times without a crew. In an endeavor that was inherently risky, the Apollo 1 disaster had made NASA risk-adverse, or at least risk-shy. Apollo 4 would sit atop the largest rocket ever built—and NASA had no intention of putting humans in the cockpit until the rocket's reliability could be vetted.

The mission of Apollo 4, besides testing the viability of the Saturn V, would be to test the stopping and restarting procedures for the S-IVB third stage—a requirement for all future lunar missions. After working through a checklist of numerous other mission requirements, Apollo 4 would complete its mission by testing the survivability of the command module heat shield upon re-entry.

The launch of Apollo 4 originally had been planned for late 1966, but the launch was postponed when a number of design and construction flaws were discovered in the S-II stage, along with additional wiring defects in the Apollo spacecraft. Once those problems were worked out, the three stages of what would become Apollo 4 were shipped to the Vehicle Assembly Building (VAB) in Florida in the spring of 1967. There they were assembled, or "stacked" as the technicians preferred to say. The rocket was assembled in its upright flight position as each stage and its separation ring was mounted one atop the other.

By late August 1967, the assembly was complete.

Dan Ruttle was one of hundreds of young NASA engineers who had been hired right out of college. Unlike most engineers, he enjoyed designing and building rockets not just for NASA, but for himself as well. Several times a year he would retire to the desert with a few other rocket enthusiasts and fly small home-built solid fuel rockets several miles into the sky. His profession was also his hobby, making rocketry an all-consuming passion.[2] Ruttle had managed to score a plum assignment as a Cape Canaveral engineer—an assignment which, two years later, allowed him to be present on the morning NASA unveiled the fully assembled Apollo 4 rocket. His vantage point was even better than that of the astronaut corps, most of whom were watching at a viewing area 3.5 miles away. He would be one of the few people to see the Apollo 4 Saturn V rocket up close once the VAB doors opened.

On the early morning of August 26, 1967, a sliver of sun cracked open the day as it rose above the Atlantic Ocean horizon east of Cape Canaveral. The launch preparations were so detailed and complex that it would be more than two months before the rocket would experience liftoff and orbital flight. Standing outside the VAB, Dan adjusted his hard hat and made small talk with some of the two dozen engineers and technicians who, like him, were waiting patiently for the largest set of doors ever built to open. Time passed slowly, and everyone kept checking their watches. As Dan checked his watch for what seemed like the fiftieth time, his attention was diverted by a warning horn that echoed across the Cape's marshy landscape. The siren dissipated, and everyone turned their attention to the VAB. The opening rumble of its massive doors was unmistakable. Both the interior and exterior of the VAB was well lit, and as the doors fully extended, the rocket appeared like a bright white beacon against the dawn sky. As an engineer Dan had always held an affinity for the brilliance and pioneering spirit of Wernher von Braun, and as he gazed at this giant creation his admiration only grew. Never before in human history had there been an attempt to build a machine the size of Apollo, let alone launch one into space.

The Saturn V stood 363 feet tall, 60 feet taller than the Statue of Liberty. Fully fueled at liftoff, it would weigh 6.2 million pounds, equivalent to the weight of about four hundred elephants. At takeoff the five clustered F-1 engines of the first stage would generate 7.5 million pounds of thrust—more power than eighty-five Hoover Dams combined.[3] With the VAB's doors opened, the massive rocket transport tractor, nicknamed "the crawler," began a ponderous forward movement, carrying the rocket into the bright Florida morn. Its movement was stunning in its lack of velocity—a sleepwalking speed of just one mile per hour. The aptly named crawler-transporter began its patient move toward the launch stand five miles away. At 131 feet long, 114 feet wide, and weighing in at 6.6 million pounds, the crawler was the largest and most powerful land vehicle ever built. It had a cargo capacity of eighteen million pounds, the equivalent of fifteen Statue of Liberties.[4] Built by Marion Power Shovel Company in Ohio, the crawler's design was heavily influenced by another product built by Marion: the giant track machines used in strip mines.[5]

Even with its fuel tanks empty, the weight of the Saturn V rocket and its transporter were immense, and NASA had designed a special heavy-duty gravel to pave the road they would travel to the never-before-used Launch Complex 39A. As Ruttle and the rest of the ground crew patiently walked beside the crawler that day, keeping an eye out for problems and glitches, he noticed an unexpected result of the crawler/rocket's massive weight: in its wake, the custom-made, high-density rock and gravel pavement the crawler traveled over was being crushed and pulverized into a dustlike substance.[6]

That this launch was occurring at all was the result of a national manned space philosophy that was veering away from long-standing beliefs held by the Saturn V's designer, Wernher von Braun. He had insisted on a methodical testing program—a proving regimen where each rocket stage was tested separately before being combined into a single unit. Having witnessed hundreds of rocket failures and explosions over the years, von Braun had burnished a reputation for careful methodology and preached a very careful and conservative schedule when it came to the progression of rocket technology, testing, and flight. However, that philosophy was defenestrated by George Mueller when he became the director of NASA's Manned Space Flight in 1963, a job he took, along with a significant pay cut, at the urging of NASA's leader, James Webb.

At the time Mueller was hired, the testing philosophies of von Braun and NASA were symbiotic—every component was tested separately, and every stage was test-fired individually. This worked fine for a while, but after the nearly two-year delay caused by the Apollo I pad fire, Mueller took a closer look at his calendar and came to a hard conclusion: Kennedy's goal of sending men to the Moon before the end of the decade would no longer be possible if the von Braun testing philosophy continued to hold sway. He decided to make a change—a big one. Prior to NASA, Mueller had worked with TRW on ballistic missile design. It was Mueller who had convinced the air force to abandon their piecemeal testing program (adopted from the Germans) and replace it with a new system he dubbed "all up." In all-up testing, the entire rocket/missile is assembled and tested as a congruent unit rather than one section at a time. The air force had proven the system's advantages, and it was based on this success that Webb had sought out Mueller to help rescue the postfire Apollo program.[7]

Apollo 4 would be the make-or-break test of the new Mueller Doctrine. Von Braun protested, but he was overruled. In the end, Mueller's all-up philosophy would be vindicated, for which he would receive a great deal of credit. But it could be argued that Mueller just got lucky—his program worked for one reason only: it benefited from the supreme engineering skills of von Braun and his team at the Marshall Space Flight Center in Huntsville.

Dan Ruttle studiously kept pace with the crawler during its four-mile, four-hour journey from the VAB to Launch Complex 39A. He had a few more duties that day, which he completed on time and without any glitches. By the late afternoon his shift was over, and he returned to his temporary digs at a local motel. There was plenty of time—the countdown to launch would take more than two more months, barring any unforeseen holds.

Most of the astronaut corps, now ballooned to forty-six men,[8] were present at Cape Canaveral on the morning of November 9, 1967, to watch the launch of Apollo 4 and the first flight of the Saturn V. Many of them would have preferred to be in the capsule that day rather than standing around like tourists. One of those rather-be-flying astronauts was Pete Conrad, veteran copilot of Geminis 5 and 11. As he stood there in the autumn chill, watching and waiting, Conrad was pensive and rueful—unlike the mucky-mucks in NASA leadership, he felt it was a complete waste of good hardware and mission goals to fly the Saturn V without a crew. Still, despite the elevated public persona that came with being an astronaut, it was a hard truth that astronauts were near the bottom of the NASA authority food chain. Like his fellow astronaut brethren, all Conrad could do that day was watch from the sidelines and hope to be assigned to some future Apollo mission.[9]

Dan Ruttle was back at Launch Complex 39A, assisting with last-minute tests and checks. Everything seemed to be working properly, and the launch countdown steadily continued. Across the bay, Dan could see the flickering pinpricks of lights—campfires strewn across the observation beach helping tourists stay warm in the chilly Atlantic Ocean air. Thousands of them had shown up to witness this historic event—some had been camped out for days.

Dan's supervisor approached and let him know it was time to go, and most of the ground crew followed him toward the launch crew's concrete blockhouse.

It is impossible to communicate in mere words a description of the Saturn V's awesome power, though hundreds of writers have attempted it. In his book *Rocket Men*, Robert Kurson says that the Saturn V produced "160 million horsepower—enough energy to power the United Kingdom at peak usage time."[10] Though I admire Kurson's book (and certainly recommend it), such descriptions, though interesting, are far too clinical. They are hardly how a launch observer would describe it. Amy Shira-Teitel gives a more "been-there" description in her Space.com article, "NASA's Gutsy First Launch of the Saturn V Moon Rocket."[11] She writes, "It shook buildings, broke windows, and its power caught many people at the Cape . . . off guard." Arthur Hill of the Houston Chronicle was at the launch, and described it from the point of view of the press box: "The powerful engines shook the press stands, rattling light fixtures and bouncing tables up and down. It was an awesome sight as brilliant yellow fire engulfed the launch pad at liftoff."[12] Gene Kranz would later write, "We sent Saturn into space on the most immense pillar of flame ever seen at the Cape."[13] The visual and aural spectacle of the rocket itself was only one part of the sensory experience of being present at the first Apollo launch. NASA's countdown communicator at the time, Jack King, said, "At liftoff, the vibration from the Saturn V showered us with dust and debris from the ceiling of the Launch Control Center which was brand new at the time." And in the broadcast booth, America's unofficial civilian space communicator, Walter Cronkite, reported that the entire building was shaking so hard that he and a CBS producer were forced to press their hands against a glass window to keep it from vibrating and breaking.[14]

Never at a loss for hyperbole, Wernher von Braun, weighed in simply: "It was really an expert launching all the way through from lifting off exactly on time to performance of every single stage," while George Mueller's comments hinted at a more positive future: "The maiden voyage of the Saturn V dramatically increased the confidence of people across the nation in the management of the largest research and development undertaking in which the western world has ever engaged."[15] In other words, with the success of Apollo 4 it looked like a manned landing on the Moon might actually be possible in the not-too-distant future.

Though all major Apollo 4 mission goals were textbook in the manner they were achieved, giving a depressed NASA organization a much-needed shot in the arm, the confidence Mueller spoke of was soon to be shaken by two more big hits—one political, and one technical.

Apollo 5 would be the first time the lunar module would be launched into space. Various aspects of the LM would be tested, including the firing of its descent and ascent engines. The descent engine would be the first throttleable engine tested in space. For the astronauts to not only land safely on the lunar surface, but take off from the Moon to return home, these engines had to be as fail-safe as possible. The most significant event planned for Apollo 5 was what the engineers referred to as the "fire in the hole test." This test would simulate a mission abort whereby the ascent engine would be fired while still attached to the descent engine—an event that hopefully would never have to happen with astronauts on board. As with Apollo 4, design and construction problems delayed the launch almost a year. LM-1, as it was designated, failed a number of crucial tests long before being sent to the Cape. In one incident, a glass window shattered during the LM's initial internal cabin pressure test—a flaw with life-threatening potential had it been manned. The decision was made to replace the glass windows with aluminum plates for the Apollo 5 mission and to work out the glass problem later.

Like Apollo 4, Apollo 5 successfully completed its most important mission goals. It launched January 22, 1968, and flew just over eleven hours. It seemed as if NASA was back in its groove.

On April 1, 1968, four days before the launch of Apollo 6, President Lyndon Johnson—the closest political friend NASA would ever have—announced he would not be seeking reelection. And if you were alive at that time and don't remember the April 6 Apollo launch, it's probably because the media was completely focused that day on a far different story: It was on April 4, 1968, that Dr. Martin Luther King was assassinated.

The last unmanned flight of the Apollo program, the main mission goal of Apollo 6 was to test the system's ability to actuate TLI—trans-lunar injection (i.e., break free of Earth's gravity). However, the thus far reliable Saturn V developed engine problems in both the second and third stages. Some of the engines in the second stage shut down prematurely, but owing to the brilliance of von Braun's quasi-religious zeal toward redundancy, the Apollo guidance system compensated for the shutdown by firing the remaining second stage engines, as well as the third stage engine, for a longer duration. As a result, the upper stage still managed to attain a workable, if not intended, orbit.

But the problems were not over. When it came time to test TLI, the third stage refused to reignite. NASA pulled out every stop in an attempt to make Apollo 6 pay for itself, but in the end the mission was aborted, and the Apollo 6 capsule was ordered into re-entry, splashing down approximately eighty kilometers from its target area.

The engine problems were traced to the "pogo effect"—a potentially dangerous oscillation that sometimes occurs in liquid fuel rocket engines, resulting in an up-and-down fluctuation in fuel flow and acceleration. In extreme cases the vibrations that result from this oscillation can tear a rocket apart. Both the Americans and the Russians experienced this problem early in their space programs, sometimes resulting in catastrophic explosions.[16]

Despite these serious problems, NASA decided to declare the Saturn V man-rated anyway—an ironic decision, given the shadow of the recently concluded Apollo 1 investigation findings.

Apollo 7 would mark NASA's return to manned space flight, and the maiden voyage of the redesigned Apollo spacecraft. It would be the first time the United States had sent three astronauts into space, and the first manned use of the Saturn 1B. It had been almost two years since the last NASA manned flight[17]—a long dry spell for a program attempting to reach the Moon by December 1969. The mission would be commanded by Walter Schirra, with Don Eisele as senior pilot and navigator, and Walter Cunningham[18] as copilot and systems engineer. Eisele had been trained as command module pilot, and Cunningham as pilot for the lunar module. All three had served as the backup crew for Apollo 1. With Grissom, White, and Chaffee gone, Schirra's crew was the next in the pecking order, and they were tapped into the flight assignment for Apollo 7.

Since the end of the Gemini program there had been a change in contractor—North American Aviation was out, and Rockwell International was in.[19] That meant a major change of ground personnel, including the launch pad leader, Guenter Wendt. Since Guenter was a North American employee, it meant he would not be working anywhere near the launch gantry any longer. Like his fellow astronauts, Schirra felt a special kinship with his German-accented friend—the last friendly face every astronaut saw as they entered their capsule. Schirra was livid that Guenter was gone, so he convinced his boss, Deke Slayton, to persuade Rockwell to hire him away from North American Aviation, and to then persuade Rockwell to place Guenter back in charge of launch pad procedures. After that, the astronauts had a new joke: "Where did Guenter Wendt? He went back." Wendt then remained as pad leader for the entire Apollo program.

Schirra would be the first astronaut to have flown all three manned programs—Mercury, Gemini, and Apollo. As NASA program managers geared up for a more demanding and progressive launch and test schedule, they decided Apollo 7 would need to help the program catch up from the post–Apollo 1 setbacks. The Apollo 7 schedule of tasks would be demanding. Among other assignments, Schirra's crew was tasked with wringing out the redesigned Apollo capsule and the fully renovated pressure suits. They would need to put them through enough tests that their viability on future missions could be assured. There would be no lunar module on board, but the crew would be maneuvering the command module in practice sessions to dock with a LM attachment mockup. The SPS (Service Propulsion System), which needed to be able to start and restart numerous times, would be thoroughly tested and, arguably most important of all, a new and improved menu of space food would be along for the ride. In addition, Apollo 7 would be the first U.S. space mission to transmit live television from space. This last item was more important than it may seem at first, as NASA was fully sold on making sure the eventual lunar landing would be adequately recorded and transmitted live back to Earth.

At eleven days, this would be one of the longest duration manned missions to date, second only to Gemini 4. Of course, NASA knew the crew might be psychologically tested themselves by being confined in such a small area for such a long period, but no one could have predicted how much that unofficial test would fail. Soon after reaching orbit, the crew began acting argumentative with each other and with their CapComs. This was later referred to in NASA lore as the first astronaut rebellion.

An example of the "rebellion" is this slice of conversation between Schirra, Jack Swigert (the CapCom), and Deke Slayton as Slayton was attempting, in vain, to convince the crew of the simplicity and importance of turning on the live television video feed.

SCHIRRA: You've added two burns to this flight schedule, and you've added a urine water dump; and we have a new vehicle up here, and I can tell you at this point TV will be delayed without any further discussion until after the rendezvous.

CAPCOM: (Jack Swigert) Roger. Copy.

SCHIRRA: Roger.

CAPCOM 1 (Deke Slayton): Apollo 7, this is CAPCOM number 1.

SCHIRRA: Roger.

CAPCOM 1: All we've agreed to do on this is flip it.

SCHIRRA: . . . with two commanders, Apollo 7.

CAPCOM 1: All we have agreed to on this particular pass is to flip the switch on. No other activity is associated with TV; I think we are still obligated to do that.

SCHIRRA: We do not have the equipment out; we have not had an opportunity to follow setting; we have not eaten at this point. At this point, I have a cold. I refuse to foul up our time lines this way.[20]

Schirra would later explain that one of the three crewmembers had brought a virus on board, and soon after launch they were all suffering from severe head colds—and their attitudes showed it. Chris Kraft displayed his wrath by making certain none of the Apollo 7 astronauts ever again flew in space.[21]

Despite his grounding, Wally Schirra would become a shining example of how hard it was to lose at anything once you had been a NASA astronaut. After returning from space and retiring from NASA, Schirra parlayed his famous Apollo 7 head cold into cash by becoming a TV spokesman for Actifed cold medication, and a member of the board of directors for Kimberly Clark—makers of Kleenex tissue.

25

MACH 33

There is no such animal as a perfect spacecraft.

—Gene Kranz

As mission commander, Frank Borman was seated on the left side of the three-man Apollo capsule. Bill Anders was on the right, and Jim Lovell was seated between them. Each of the astronauts took occasional glances at the mission clock as it ticked steadily downward. Soon they would attempt something no living creature had ever done: escape Earth's gravitational hold. The men and their vessel were gliding in orbit, but despite their spacefaring state they were still in gravitational capture, like a dog on a leash circling its owner. If all went according to plan the astronauts would soon take an axe to that leash, hurling themselves out of Earth's orbit and severing their home planet's gravitational vise grip. It would be an historic event of mammoth proportions, and everyone involved knew it.

The mission was named Apollo 8, and the three men assigned to it had been chosen due to their wide range of abilities and talents, not the least of which was their businesslike no-nonsense attitudes—especially Mr. Straight Shooter himself, Frank Borman.

Of all the newly established traditions in the space program, the one that arguably should have been done away with by the time of the Apollo 8 mission was the ceremonial steak-and-eggs breakfast just prior to launch. As the men unbuckled themselves from their seats one by one, each felt the pangs of nausea and the urge to vomit. The Gemini capsule had been too cramped to allow in-capsule movement, but now with the much larger Apollo capsule, the men could get out of their seats and move around, enjoying the benefits—and detriments—of weightlessness. And nausea was one of the detriments. All three men would suffer some discomfort, but it was Borman who was in for the worst of it.[1]

Space sickness or not, they were busy—NASA had given them only two Earth orbits in which to complete a myriad of preparatory tasks. They had practiced and drilled these tasks many times in Earth-bound simulators, none of which were capable of perfectly replicating the zero-g environment they now found themselves in. While they worked, the men kept up an active communications link with Mission Control—mostly of the routine "Everything looks good here" sort. Thus far, there had been no major glitches. The launch had been flawless, the stage separations had worked according to design, and the installation of the Saturn S-IVB third-stage-plus-capsule spacecraft into orbit also seemed perfect.

But the next step would be the big one: to reignite the third stage engine, break free from Mother Earth's persistent gravitational well, and fly toward the Moon. This maneuver was referred to in NASA-speak as TLI, for trans-lunar injection. The boyhood dreams of Robert Goddard, Wernher von Braun, and Sergei Korolev were coalescing into reality so spectacular that half the people on Earth would be transfixed to television screens for at least some portion of the mission.

Apollo 8 would not land on the Moon, but its mission would be the greatest undertaking thus far in human history: to break free from Earth orbit, fly 240,000 miles to the Moon, orbit it ten times and then, hopefully, return its three crewmen safely to Earth.

Escape velocity—the speed that must be reached to unleash an object from Earth's gravity—was a well-known value. It was based on the math and natural laws first outlined by Sir Isaac Newton three hundred years before Apollo and by Konstantin Tsiolkovsky a century before. For planet Earth, that value was 25,020 mph—what is sometimes approximated as Mach 33 (i.e., thirty-three times the speed of sound). It was often reported in the media that Apollo 8 would reach escape velocity, but in truth, it would intentionally miss that mark by a small margin. As a safety precaution, the engineers at NASA had programmed the mission escape speed to 24,200 mph—a speed fast enough to get them to the Moon, yet slow enough to return the ship to Earth by gravity if the mission had to be aborted.[2] This decision came with a secondary benefit—it would cause the Apollo spacecraft to gradually slow down as it pulled away from Earth, meaning less impulse from the engine would be required to slow the craft down for LOI (lunar orbit insertion).

For Apollo 8, one of the main CapComs would be Michael Collins. Chosen as a member of the third group of fourteen astronauts in 1963, Collins was also a former test pilot, and a veteran of Gemini 10. As NASA flight controllers pored over the data streaming down from the command module, all flight systems appeared to be operating within required parameters, and so the decision was made to allow Apollo 8 to fire its engine at the appointed time for TLI. Frank Borman and his crew were given the green light to break free of Earth and head to the Moon. Flight's decision was handed down to Collins, who sent the good news up to the anxiously waiting astronauts.

"All right, Apollo 8. You are go for TLI."

Though trans-lunar injection would be an important scientific and historical milestone, it was a political one as well. Until now the space race between the United States and the Soviet Union had been somewhat neck-and-neck, with each country occasionally leapfrogging the other. The Soviet Union had started the race with Sputnik and lengthened their lead with Luna 3's photography of the back side of the Moon and Gagarin's pioneering flight. The Americans had fought hard to catch up—and were succeeding. But TLI would change everything about this undeclared contest. Races are emotional and philosophical rather than scientific, and it was the same with the Space Race. There was no scientific or intellectual marker in the idea that the Americans and the Russians were "racing" in their goals to exploit space—no starting line, no finish line, no agreed-upon method of determining a winner. Yet it was the mere idea that such a race existed that forged the emotional zeal necessary to keep space progress moving at a torrid pace for both countries. TLI would change everything because it would greatly alter this philosophical duel. The moment the astronauts broke free of the Earth and headed to the Moon would mark a milestone so extraordinary as to make the United States's place in that race preeminent. After the world's first TLI, the United States would never look over its shoulder again.[3]

Mike Collins was one of the few who understood the breakaway nature of this event. In his seat at CapCom, Collins reflected on the unheralded moment. He felt there should be some celebratory public display of acknowledgment—fireworks, or "an oompah band."[4] Instead there was nothing but the dull jargon of, "You are go for TLI."

TLI was now just five minutes away. In the capsule, the men prepared for the moment by checking their instruments, loosely fastening their safety belts (with the low g acceleration to follow, tightened belts would not be needed), followed by rechecking their instruments yet again. Then, with all preparations completed, all that was left to do was watch the clock ticking downward. At T-minus ten seconds, Borman began a verbal count. And when he hit "zero," the men felt the jolt of acceleration as the hydrogen and oxygen propellants combined in the engine's combustion chamber, ignited, and pushed their craft to higher speeds.

Once again, their eyes roved over their instruments to make certain all was going well with their spacecraft. The men called out instrument figures to each other for verification, and Collins radioed up occasional messages of encouragement. The burn would last for precisely five minutes, twenty seconds. The culmination of the firing had been dubbed by NASA as SECO, an acronym for "sustainer engine cutoff." During the firing the astronauts passed the time with as much aplomb as could be expected. The onboard clocks ticked onward, and then it was over. At the appointed moment, the onboard computer system did its job perfectly, shutting down the engine, initiating the craft's lunar-targeted nonpowered coast.

"Okay, we got SECO right on the money," announced Borman.

"Roger," replied Collins. "Understand SECO."

It took less than five minutes for Apollo 8 to beat the Gemini 11 altitude record of 850 miles. After SECO, at 24,200 miles per hour, Frank Borman, James Lovell, and Bill Anders were the fastest humans who ever lived (if you didn't count the fact that the Earth itself, and its inhabitants, revolve around the sun at a speed of 67,000 mph).

The men were well aware of how urgent their job was. The only way President Kennedy's promise would happen was if Apollo 8 succeeded. Just as the Apollo 1 fire had created an eighteen-month delay for investigations, hearings, and product redesigns, any mishap or major mission failure now would undoubtedly push the program back even further. "Before this decade is out" would become "sometime next decade." This was why Borman had been tapped to be the mission commander—NASA was very aware of how sober, serious, and no-nonsense a leader he was. Apollo 8 had to succeed, and no one was more qualified to lead it.

After SECO, the next major event was to disconnect the command module from the Saturn S-IVB third stage. Its fuel was mostly spent; the rocket was now dead weight. A series of computer commands had to be entered for the procedure—a job assigned to Lovell. Once that was done, the computer asked a simple question: PROCEED? Lovell entered the command for YES, and a ring of explosive bolts fired, disconnecting their craft from the now useless S-IVB. Borman, as pilot, grabbed the controls and gently fired a short burst of his thrusters—pushing the craft a safe distance from the tumbling spent stage. He then maneuvered the craft around 180 degrees in order to get a view of the S-IVB and make certain there was a sufficient gap between it and their spacecraft.

And that's when they saw the Earth—from farther away than anyone had ever seen it.

Anders, in his capacity as crew photographer, snapped several photos of both the Earth and the retreating S-IVB. He would have continued in that vein had not Borman stopped him. Though he knew Mission Control wanted such pictures, Borman was more focused on the mission.

The coasting portion of the mission had begun. Lovell removed a tool from its storage spot to take measurements of their craft's position. The tool was decidedly low-tech, but in 1968 it was still state-of-the-art. That tool was a sextant, the same tool used by ancient mariners, now transported aboard the most high-tech machine ever built. Using several stars as reference, Lovell would use it during the mission to verify the trueness of their position and trajectory.

Throughout the coasting portion of the flight there was still much work to do, but an unexpected problem soon developed that the men were unwilling to announce over a hot mic, lest everyone at Mission Control hear and know about: Frank Borman's initial motion sickness and nausea had degenerated into something much worse. He had developed a gastrointestinal problem involving vomiting and diarrhea. At Borman's insistence, his condition was kept secret for almost twenty-four hours, before Mission Control finally figured it out on their own by downloading and listening to the tape recordings of the private conversations between the astronauts.

Dr. Charles Berry—the chief Mission Control physician—was summoned. What he heard on the tapes alarmed him. For years NASA had been concerned about the possibility that astronauts could become sick by passing through the Van Allen radiation belts—an event that had occurred for Apollo 8 at the 620-mile altitude point. But as minutes and hours progressed, and neither Lovell or Anders showed any symptoms, the doctor eliminated radiation as the cause. Instead his conclusion was that Borman had brought with him a virus, which meant Lovell and Anders could catch it. Berry suggested to Mission Control that they consider cancelling the mission. But when his suggestion was relayed up to the astronauts, they refused to considerate it. Borman called the idea "unadulterated horseshit."[5]

With that, the mission was full on. For the next few days all that would be necessary would be to wait for the ship to arrive in lunar orbit—which velocity and inertia would reliably take care of. There would be a couple of mid-course corrections, numerous instrument and equipment double-checks, occasional urine dumps into space, and—when the men allowed themselves the luxury—the occasional nap. Like Wally Schirra in Apollo 7, Borman bristled at NASA's requirement that all three astronauts take their sleep time simultaneously. Like Schirra before him, Borman felt simultaneous sleeping was negligent—that at least one astronaut should be awake and on duty at all times. As mission commander, he decided to follow his gut and ordered the two-off, one-on sleep schedule that would last most of the mission.

At about two hundred thousand miles altitude, the gravitational pull of the Moon became greater than that of the Earth, resulting in a gradual increase in their craft's velocity. As the command module approached the Moon, its speed increased from 3,800 to 5,800 mph.[6] The ship, essentially flying backward with its rocket nozzle pointed in the direction it was going, would fire a long four minutes, two seconds. The astronauts spent much of their time preparing for that crucial burn.

Sixty-eight hours and four minutes into their historic flight, CapCom sent up word to Apollo 8 that they were approved for the maneuver known as LOI: Lunar Orbital Insertion. Astronaut Jerry Carr was manning CapCom at the critical moment.

"Apollo 8, Houston. At sixty-eight-oh-four you are go for LOI."

The firing would occur shortly after the Moon's gravity bent the craft's trajectory around its mass, meaning they would be behind the Moon at the point of ignition. Mission Control therefore could neither control nor monitor any of it—Apollo 8's radio and telemetry would be completely cut off from Earth for an agonizingly long thirty-five minutes. Only after re-emerging from the Moon's far side—assuming it did emerge—would Mission Control know the crew's fate. A few seconds before radio blackout, Carr radioed the crew one last message.

"We'll see you on the other side."

Until now Apollo 8 had spent much of its voyage flying backward, its nose pointed toward Earth. As a result, the crew had had no opportunity to see their lunar target—essentially flying on faith. Now, as they coasted gradually into the Moon's gravitational well, Lovell looked out his window and had his first close-up. He and Anders could not contain their enthusiasm as they watched craters, valleys, and rills unspool beneath their position.

Borman got them back on task, and at the appointed moment ordered the required retro-burn of the engine, slowing the craft down to orbital speed. Once again, the engine operated flawlessly. It was December 24, 1968, and with Earth out of radio contact, only the three-man crew were aware everything was "A-OK."[7]

The command module and crew of Apollo 8 continued to score more aeronautical firsts, not the least of which was the fact they were the first humans to see the far side of the Moon with their own eyes. To Borman's discomfort, Anders could not stop taking pictures, and Lovell could not resist sightseeing, as both men floated from one window to the next. Meanwhile, back on Earth, three billion people anxiously awaited word. Was Apollo 8 in a proper lunar orbit, or had something gone wrong? If the braking rockets fired too long, the craft could crash into the Moon—if fired not long enough, they would shoot past the Moon and into deep space. Of all Earth's population, no one, of course, was more anxious than the crew's families, each of whom sat patiently next to their "squawk box"—speakers set up inside their homes by NASA through which the families could monitor communications between the crew and Mission Control.

To the credit of the mathematicians, the computer programmers, the hardware designers, and the crew, Mission Control reacquired Apollo 8's telemetry signal within one second of when the math said it should.

Carr hailed the ship. "Apollo 8, Houston. Over."

"Go ahead, Houston. This is Apollo 8."

Whatever was said next was drowned out in a huge outpouring of whoops and cheers throughout Mission Control.

Borman and his crew were scheduled to make ten complete orbits around the Moon before re-igniting their engine and heading back to Earth. During that time NASA had assigned them a myriad of experiments and assignments. As with the trip from Earth to Moon, their sleep schedule would be constantly shortened by their zeal to get as much accomplished as possible. As a nod of thanks to those who got them there, the crew named craters and other landmarks after them—including astronauts who had passed away in tragic accidents. Lovell even found a peak he named after his wife: Mount Marilyn.

Earthbound humanity participated in the mission by tuning into one of a half-dozen scheduled broadcasts from the ship. Borman disdainfully labeled them "dog and pony shows," but NASA (and the public) insisted on them. Three of those had already come and gone. Of the three yet to come, none was more anticipated than the one scheduled for Christmas Eve, shortly before the crew was scheduled to leave lunar orbit. The subject of the Christmas Eve broadcast had been a matter of debate for months. The only thing everyone was certain of was that it should be "something special." Three weeks before liftoff Julian Scheer, chief public affairs officer in Houston, told Borman, "You're going to have the largest audience that's ever listened to or seen a television picture of a human being. And you've got, I don't know, five or six minutes."[8]

When Borman asked what they should do with that time, Scheer replied, "Do whatever's appropriate."

Borman considered their options carefully, but could not come up with a satisfying idea. When he brought the matter up with Lovell and Anders, they also drew a blank. The matter was handed off to numerous individuals in various departments, whose lack of ideas illustrated how much better NASA was at building rockets than producing television. Joe Laitin, a former public affairs officer for President Kennedy, was one of those grappling with the problem. He mentioned it to his wife Christine, who also had no suggestions. At around 4:00 a.m. a few nights later, Laitin was in his kitchen still working the problem when Christine walked in. She had an idea and presented it to him. Right away, Laitin understood she had got it right.

Her idea—like all great ideas—was the epitome of simplicity. A few days before the scheduled launch, Laitin sent it up the chain of command to the astronauts, who also agreed it was appropriate.[9]

As the astronauts pulled around from the far side of the Moon on their next-to-last orbit and faced Earth and its inhabitants, the astronauts began their televised Christmas Eve program. At first it was very simple—each man taking turns attempting to describe, within the natural limits of human language, what it was like to fly over and see up close Earth's lonesome satellite.

"The Moon is a different thing to each one of us," began Borman. "It is a vast, lonely, forbidding-type existence or expanse of nothing. It looks like clouds and clouds of pumice stone."

"The vast loneliness up here of the Moon is awe-inspiring," added Lovell. "And it makes you realize just what you have back there on Earth. The Earth from here is a grand oasis in the vastness of space."

Anders added, "I think the one thing that impressed me most were the lunar sunrises and sunsets."

The three poets-of-the-capsule spent a few more minutes describing to their audience what they were seeing, then Borman announced, "For all the people back on Earth, the crew of Apollo 8 has a message we would like to send you."[10]

Despite the high-tech nature of Apollo 8's mission, their flight plan was still printed on paper, albeit a special fireproof type. In the last page of the flight plan had been inserted their Christmas Eve message. Anders now opened to that page.

"In the beginning," he read, "God created the heaven and the Earth."

Taking turns, each astronaut read a passage of the first ten verses of Genesis, often referred to as the "creation story." The publicly aired reading of these verses put Apollo 8 and its magnificent mission in an entirely new perspective.

Even the stodgy science geeks at Mission Control were moved by it. An estimated one billion people on Earth had been watching and listening in, and the effect on humanity was electric. When the reading was done, Borman signed off with a humble close.

"And from the crew of Apollo 8, we close with good night, good luck, a merry Christmas, and God bless all of you, all of you on the good Earth."[11]

Even before the mission was over, pundits, journalists, politicians, and scientists from around the world began weighing in on the significance of Apollo 8's many achievements. Sir Bernard Lovell (no relation to James Lovell) of the Jodrell Bank Observatory in Great Britain described Apollo 8's success as "one of the most historic developments in the history of the human race." In the Soviet Union, Boris Petrov, chairman of the Soviet Interkosmos program, described the flight as an "outstanding achievement of American space sciences and technology."

Borman and his crew would complete all ten of their planned orbits around the Moon before once again firing their engine to break lunar orbit and return to Earth. They would make a near bulls-eye splashdown in the northern Pacific.

Michael Collins—future command module pilot of Apollo 11 and the first Moon landing mission—would one day refer to Apollo 8's successful mission as more significant than the Moon landing itself.

26

AMIABLE STRANGERS

Buzz was generally quiet and incapable of small talk.

—Michael Collins

Neil Armstrong was the best pilot I had ever met.

—Buzz Aldrin

Armstrong was content that Mike Collins and Buzz Aldrin would serve with him on Apollo 11.

—James R. Hansen

Two weeks after the success of Apollo 8, and the worldwide adulation it brought to the United States, NASA had yet to announce who the crew would be for the first lunar landing, tentatively planned for either Apollo 11 or Apollo 12. Every member of the astronaut corps had lobbied to one degree or another for the right to be one of the three chosen for what would be the most historic voyage in human history. Whispers and rumors were rampant throughout NASA, among the journalists who covered the space program, and among millions of average Americans everywhere. Who would be the men to be the first to land on the Moon?

Secretly, von Braun, James Webb, Chris Kraft, and dozens of others in space's inner circle were still fretting that the Soviet Union would announce any day that their country had just landed cosmonauts on the Moon, notching yet another record for the books. Soviet firsts in space had happened so often in the past that paranoia had seeped into NASA like a virus. Everyone from controllers to astronauts, technicians to engineers, continued to second-guess themselves. Yes, Apollo 8 had buoyed their enthusiasm, but the Central Intelligence Agency kept producing spy plane flyover photographs of a giant rocket being assembled in Baikonur, a rocket as large as the Saturn V.[1]

On July 4, 1969, the world discovered at least one of the intended missions of this new Soviet rocket. As if intentionally throwing fire on NASA's paranoia (on America's Independence Day, no less), the Soviet Union launched two Soyuz spacecraft into orbit. The two craft rendezvoused after only two orbits, then the two crews achieved something the Americans had barely even considered doing: they transferred two crew members via EVA from one capsule to the other, then both capsules returned to Earth. And thus another Soviet space first was officially in the record book.

Five days later, NASA held a press conference to announce the crews of Apollo 11 and Apollo 12. The Moon landing mission was officially announced as Apollo 11, and the crew would be Neil Armstrong as mission commander, Buzz Aldrin as lunar module pilot, and Mike Collins as command module pilot.

A contrast between the Apollo 11 and Apollo 12 crews and their families soon became apparent. Pete Conrad, Alan Bean, and Dick Gordon had many things in common, not the least of which was their background as navy pilots. The men drove matching gold Chevy Corvettes, and their wives took every opportunity to socialize together and bond, even to the point of wearing matching outfits.[2] The Apollo 15 crew also drove matching Corvettes (red, white, and blue), and the Apollo 10 crew, led by Gene Cernan, were close friends who were proud of their ability to work well together as a team. They stayed together after their workday was done, having a beer (or two, or three), telling jokes, finding ways to blow off steam, and generally have fun together in the few off-work hours NASA allowed.

The Apollo 11 crew fit a different mold.

The three astronauts formed an unlikely trio. Whereas most of the members of the astronaut corps were gregarious and extroverted, Armstrong, Aldrin, and Collins each had a "loner" reputation. They kept to themselves more than the others, and were often lost in their own worlds. Mike Collins would later refer to himself and his two crewmates as "amiable strangers." The same applied to their wives, who rarely socialized, coming to together only when required for NASA activities, or for the Astronaut Wives Club.

When their work was done each day, Collins, Armstrong, and Aldrin would simply go home—no late-night bar visits or Corvette races through Cocoa Beach. The rousing and carousing lifestyle of their astronaut corps comrades was not in their nature. Was it happenstance, divine intervention, or simple NASA calculation that the three ended up together? In the case of Armstrong, no member of the astronaut corps was more surprised than he when Deke Slayton tapped him to be an Apollo mission commander. Armstrong was a humble, soft-spoken, highly intelligent engineer and pilot, and one of the few that everyone in NASA seemed to admire. But many NASA people considered his soft-spoken demeanor a liability in an astronaut corps filled with gregarious and colorful characters. How could someone stand out from such a crowd? Yet somehow Armstrong managed to do it.

Aldrin seemed a logical choice for the first lunar landing mission; for several months Slayton had seriously considered putting Jim Lovell in that seat, and had told Armstrong as much. But Aldrin had several things going for him that would put him head and shoulders above any other choice. He was the only member of the astronaut corps with an engineering PhD, and that degree was in a field of crucial importance: space rendezvous. His knowledge on the subject of orbital flight mechanics and a dozen other crucial performance subjects was legendary. The only real surprise was Collins.

Michael Collins had actually been removed from active flight status in August 1968 due to an unexpected medical problem: "There was something wrong with me, something insidious, something getting worse, something obviously serious."[3]

During a game of handball, Collins noticed that his legs were not following the commands his brain was sending them. At first he dismissed the symptoms, chalking it up to fatigue from his rigorous astronaut training. He was having trouble running, going downstairs. Sometimes a simple level walk gave him problems. One day, while descending a flight of stairs, his left knee spontaneously buckled, and he almost took a tumble. Thereafter he began feeling a tingling sensation in his left leg, and parts of the leg would occasionally go completely numb with no feeling at all. After a few more days, as the numbness and tingling began to travel up toward his thigh, Collins decided to do the one thing all test pilots and astronauts avoided like the plague: he would visit the flight surgeon. He was fully aware, of course, that seeking help might get him kicked off active flight duty. Collins had already been bumped from one crew in favor of Jim Lovell—the crew that became the historic Apollo 8 mission.

On July 12, 1968, Collins walked into the NASA flight surgeon's office at the Cape. The doctor was unable to diagnose Collins, who was sent to see a specialist in Houston where a series of X-rays were taken. The diagnosis: a bony growth between his fifth and sixth cervical (neck) vertebrae that was pushing against his spinal cord. Immediate surgery would be needed.[4]

The operation at the air force's Wilford Hall Hospital in San Antonio was a success, but the convalescence would take several months. And he was removed from active flight status. Would Mike Collins ever fly an Apollo mission? It was a question that would keep him up at night. Yet almost miraculously Collins's health improved so fast that by the end of November, he was restored to full flight status.

Collins busied himself in Houston, giving speeches and acting as a NASA spokesman whenever called on to do so. He also advocated and assisted the Apollo 8 crew as they readied for their historic journey around the Moon. He was rewarded for this work by being assigned as the Apollo 8 CapCom on the Green Team—the team assigned for launch operations.

A week later, with the splashdown of Apollo 8 on December 27, 1968, it was clear that America had finally passed the Soviet Union in space exploration. Thirteen days later, on January 9, 1969, NASA announced that Apollo 11 would be the first lunar landing mission, and that it would be crewed by Neil Armstrong, Buzz Aldrin, and Mike Collins.

Of his inclusion with Armstrong and Aldrin as an Apollo 11 crew member, Collins would write in his book, *Carrying the Fire*, "I considered myself damned fortunate to be joining them."[5]

27

DRESS REHEARSALS

One after another the major impediments to a lunar landing were being removed.

—Michael Collins

One of the challenges of going "where no man has gone before" was coming to a consensus and agreement over flight missions, trajectories, goals, methods, and a million other details. As it became more apparent that a Moon landing might actually happen, a group of about sixty NASA managers, engineers, and mission controllers gathered in what was billed as the Lunar Landing Panel. The purpose of this panel was to discuss and decide on the tasks the astronauts needed to complete once they left the command service module and entered the lunar module. What had to happen first? What had to happen second? What needed to be accomplished the first hour? The second hour? And so forth. The meeting was held to garner a wide selection of ideas from the movers and shakers within NASA in order to find the answers to these questions. They would spitball hundreds of ideas and glean the wheat from the chaff. Steve Bales, a mission controller who had wormed his way into NASA before he even graduated from college, and who attended the Lunar Landing Panel, described the meeting as "a total cacophony of opinions. People [were] talking at once, sometimes shouting. Once in a while someone would get so mad they would have to leave the room."[1] It took half a day to come up with a decision on what the tasks would be for just the first hour of LM operation. Despite the stock exchange–like brouhaha, after two days the panel managed to come up with a tentative outline on initial LM operating procedures.

Once Apollo 8 proved that navigation to and from the Moon could be safely executed, it was time for the dress rehearsals to begin, and those rehearsals were Apollos 9 and 10. These missions would test every major aspect of the spacecraft hardware and computer software in preparation for opening night—the Apollo 11 landing.

Launched on March 3, 1969, Apollo 9 would be the third crewed Apollo mission and the first to carry a working LM. For the first time, the CSM (named Gumdrop) and the LM (named Spider) would be flown together, separated, then maneuvered and docked with each other. Of the two docking maneuvers, one was with the LM still mounted inside the S-IVB stage, and the second was with the LM flying free of the S-IVB. In addition, the crew would test transferring astronauts between the CSM and the docked LM in real-world weightless conditions.

The mission commander was James McDivitt, David Scott was the command module pilot, and Russell Schweickart acted as lunar module pilot.[2] An important goal of the mission was to practice the many engine ignitions, cutoffs, and throttlings that would be used in a lunar landing mission. The many engine firings that occurred during the missions, some of them very lengthy, created a flight pattern that almost seemed like gymnastics. Their orbital speeds, altitudes, and shapes changed frequently over the ten-day mission, proving how versatile and maneuverable the two ships were.

While its immediate predecessor, Apollo 8, had gained a great deal of glamour from its Earth-Moon voyage, it was Apollo 9 that worked out the nuts and bolts and maneuvers of what would one day be a lunar landing.

To illustrate how incredibly accurate NASA's flight plans and calculations had become, the Apollo 9 astronauts splashed down in the Atlantic Ocean 241 hours and 53 seconds after launch—only 10 seconds off the preflight plan.[3]

Once it had been proven that the LM's systems functioned reliably in space, the next step would be to have it fly close to the Moon (but not land). The mission of Apollo 10 would be a lunar landing dress rehearsal. Commanded by Thomas Stafford, the CSM pilot was John Young and the LM pilot Eugene Cernan, all veterans of Gemini missions. The astronauts decided to again give their spacecraft whimsical names; the CSM was named Charlie Brown, and the LM was dubbed Snoopy, after the Charles Shultz cartoon characters popular at the time. They also carried aboard the first spacefaring color television camera—something that made the TV networks much happier.

Launched May 18, 1969, Apollo 10's mission plan was to fly the LM down to an altitude of 15.6 kilometers (about 9 miles)—the point where powered descent would occur on future landing missions. NASA designed this mission as an Apollo 11 dress rehearsal not only for the astronauts and their equipment, but for the mission controllers as well.

Apollo 10 would also help answer questions the flight engineers had about anomalies in lunar gravity. Astronomers had recently discovered, verified by Apollo 8, that the Moon had "lumpy gravity"—its mass was unequally distributed. Some sections of the orb were more massive than others. These areas of greater density were called mascons (short for mass concentrations), and a big question that had to be answered was how these would affect lunar trajectories—especially landing trajectories. According to Jay Melosh, a geophysicist at Purdue University, "Mascons were . . . gravitational hazards. . . . They were a real pain in the neck for Apollo planners—like reefs in an ocean, they were things to be avoided and planned around."[4]

NASA—ever mindful of the rebellious nature of their test pilot astronaut corps—was so concerned that Cernan and Stafford might go off-book and attempt a landing of their LM that NASA intentionally underfilled the LM's propellant tanks, thus making a landing impossible. As John Young remained with the CSM at an orbit of 70.5 miles, Stafford and Cernan took the LM down to its planned low-altitude approach, cruised there for a short while, took reconnaissance photos of the Apollo 11 landing area at the Sea of Tranquility, then returned and docked with the CSM. According to NASA, "All mission objectives were achieved."[5]

28

THE WAR OF "WHAT IFS"

For space flight a good simulator is an absolute must.

—Michael Collins

Astronaut flight training for the Apollo missions consisted of two forms—real-world machinery (planes, jets, helicopters, etc.) and "black box" computer-controlled environments that simulated what the astronauts would see and experience during their flights. Creating simulations to rehearse space flight was crucial to proper training. The goal was to anticipate all the "what ifs"—what if the lunar module loses altitude control? What if the command service module burn does not shut off at the proper moment? What if the lunar approach speed is too high? There were thousands of these "what ifs," and the simulations were an attempt to prepare for as many as possible.

As part of the astronaut training, Deke Slayton encouraged the astronauts to continue to hone their flying skills. According to Slayton, "The airplane is our only dynamic simulator. If you make a mistake you've got to get out of it, or solve the problem. You can't reset the computer."[1] To Slayton, real-word flying gave the trainee something a computer simulator could never offer: the adrenaline rush of actual risk. It was for this reason that whenever the astronauts had to travel long distances, such as for public relations assignments or meetings with aerospace contractors, they often flew T-38 trainers to get there. This policy, of course, had catastrophic results in the cases of Ted Freeman, Elliot See, and Charles Bassett, all of whom died in T-38 crashes while engaged in their astronaut duties.

CHAPTER 28

With the exception of geologist Harrison Schmitt, all the Apollo astronauts had been fixed-wing pilots. Very few of them, however, had ever piloted a helicopter, a machine whose skills replicated some of those that would be needed with the LM. So early on in the program, Deke Slayton decreed that all astronauts should have at least some helicopter hours, and a new regimen of training was added to an already busy schedule.

In addition to helicopter training, the astronauts also had at their disposal a spider-like flying machine called the LLRV—Lunar Landing Research Vehicle.[2] Built by Bell Aircraft with help from NASA, the LLRV used a downward-pointing General Electric 4,200-pound thrust jet engine to simulate a one-sixth gravity, Moon-like environment.[3] Like the LM that would land on the Moon, the LLRV was a gangly-looking asymmetrical flying contraption with four legs. Mike Collins described it as a "weird-looking, flying-bedstead contraption."[4] All LM crew members, be they LM pilots or not, were required to have at least a few turns at its controls. A pilot would sit in its cockpit and practice vertical takeoffs to around 1,500 feet before executing a gentle landing.[5] At least that was the intention. In reality, the LLRV proved to be more difficult to fly than the LM, making it a poor simulator.

Neil Armstrong almost lost his life in one.

On May 6, 1968, fourteen months before he was scheduled to land on the Moon, Armstrong was in the LLRV at Ellington Air Force Base near Houston on a routine training flight. Everything seemed to be going well until a leak in the helium line that controlled the steering jets caused a failure of the LLRV's flight control systems, and it began to list backward.[6] This caused the jet engine to point farther and farther from its crucial vertical flight position—a dangerous situation. Had the engine gone as far as pointing horizontal to the ground, lift would have vanished, and the machine would fall like a rock. As the tilt approached the horizontal level, Armstrong actuated the ejector controls, launching himself three hundred feet into the air. The LLRV crashed two seconds later in a fireball. Armstrong, uninjured save for a bit tongue, gently floated to Earth nearby. An hour later, to the great surprise of fellow astronaut Alan Bean and others, unflappable Neil Armstrong was back at work in his office.[7]

After the real-world adrenaline machines, there were the far safer, black-box computer-controlled flight simulators. When the Apollo astronauts were first introduced to the simulators, they were amazed at the leap in technology and complexity—they made the Mercury program simulation machines look like toys. Each Apollo simulator was designed to replicate the capsule in its design, layout, electronic readouts, controls, and even its cramped quarters. According to Mike Collins, "There were over three hundred of one type of switch alone, not to mention the scores of pipes, valves, levers, brackets, knobs, dials, handles, etc., etc."[8] The goal was to make the simulated training experience as close to the real thing as possible.

Each simulator had a tech team assigned to it, with a simulator supervisor in charge. In classic NASA-style name shortening, these leaders came to be known as the SimSups. The purpose of the simulation team was to test the knowledge and psychological readiness of the astronauts and their controllers to see how well they understood the Apollo hardware, its operation, and the completion of the assigned flight plan. The team accomplished these tests by becoming "monkey-wrenchers"—simulating as many challenging and crazy problems as possible for the astronauts and mission controllers to solve, thereby throwing a wrench into their confidence and reserve. The astronauts and the mission controllers were tasked with finding solutions to those problems, so that the mission and flight goals were carried out.

In theory, the purpose of the flight simulation training exercises was simple: to professionally prepare the astronauts for the unexpected problems that were sure to happen in space. In practice, however, the Apollo simulations evolved into fierce rivalries. First there was the rivalry among astronaut mission teams to have preferred access to the simulators. There were only two CSM simulators at Cape Canaveral, and at least five astronauts competing daily for its use.[9] Whichever crew was next in the flight order had absolute carte blanche priority over the CSM and LM simulators—they were at the top of the pecking order and could push everyone else aside. The astronauts scheduled for later flights then fought over the remaining crumbs. It was supposed to be orderly, but arguments inevitably erupted.

The second rivalry was "crew versus geek." Each mission crew of three astronauts would undergo simulator training assisted by means of a voice loop by the technicians and engineers from Mission Control, who were tasked with guiding the crew through glitches, bugs, and anomalies. Challenging the crew was the SimSup and his team of tech-savvy computer gurus tasked with inventing those glitches, bugs, and anomalies and programming these into the simulators. Each side was highly competitive and very focused on "winning." For the simulation teams, winning meant finding problems the astronauts and controllers could not resolve, or at least not resolve in a time period that prevented a "loss of mission goals" (a euphemism for an explosion or crash). For the astronauts and controllers, winning meant resolving every problem thrown at them by the simulators, and doing so in a timely mission-completed manner.

It was important to make the simulators mimic reality as much as possible. Whenever a pilot was recruited into the astronaut corps, one of the many surprises they would meet was the quantum leap in complexity and realism of the space flight simulators compared to the spartan plywood black boxes they were used to at the Edwards or Patuxent River facilities. According to Mike Collins, "The first time John Young saw the CSM simulator, he dubbed it the Great Train Wreck, in tribute to its physical bulk and geometric complexity. It was a huge and gaudy affair, with a carpeted staircase leading up to the cockpit entrance, some fifteen feet above floor level."[10]

The simulators were run by massive punch-card-fed computers costing millions of dollars, attended to seven days a week by hundreds of people working three shifts around the clock. Despite all that money and effort, in the early days of the Apollo program it was often easier to make the rockets fly than to get the simulators to replicate the flying.[11]

Eventually the computer engineers and technicians were able to get most of the kinks out of the system, and the simulations became more reliable and closer to the real thing. When Borman, Lovell, and Anders began simulating their orbital lunar missions, the realism of the simulations even extended to the three-second time delay between Earth and Moon. By the time Apollo 8 parked itself into lunar orbit, the astronauts would be a quarter-million miles from Earth. After their radio signals were beamed to Earth, then bounced around the globe until they hit Houston, there would be a three-second time delay between what Frank Borman said, and what Mission Control heard. Radio signals from Earth to Moon would suffer the same problem. This meant the actual communication delay was six seconds at minimum—three seconds down, and three seconds back up for the answering of any questions. Add to that the time required for Mission Control to analyze a problem and figure out a solution before sending it back, and it was obvious that any situation that required a prompt response was in trouble. To adequately rehearse for this problem, the communication part of the simulators was designed with time-delay circuitry to artificially replicate the delay between the astronauts and their controllers, even though they were only hundreds of feet apart.

Even Gene Kranz, always Mr. Frustration when it came to hiccups, delays, and mistakes, got to the point where he was happy with the state of the simulations: "[By the] late 1960s our simulation technology had progressed to the point where it became virtually impossible to separate the training from the actual missions. The simulations became full dress rehearsals for the missions down to the smallest detail."[12]

Of all the Apollo flight simulations for which the astronauts would be tested, none was more crucial than lunar powered descent. Most of the rendezvous maneuvers had already been practiced with the Gemini and early Apollo missions, but the actual landing on the Moon would be a first. Dick Koos, a former sergeant with the Army Missile Command at Fort Bliss, was the SimSup in charge of the lunar landing simulations. He had earned this vital responsibility as a result of his expertise in the computer guidance of ground-to-air missiles. Flight simulation was new for him, but he applied himself and soon became the simulation go-to guy for the Apollo program. According to Kranz, Dick Koos was "like a rapier, cutting so cleanly that you did not know you were bleeding until long after the thrust. Koos was a worthy adversary."[13] During one particular training exercise, Koos would throw an error code at the astronauts and MCC that would become a footnote in space history.

As the launch of Apollo 11 approached, Armstrong and Aldrin gobbled up the lion's share of time on the LM simulator. Even so, the next-to-fly team of Pete Conrad and Alan Bean in Apollo 12 managed to get in some simulation time as well. Koos and the geek monkey-wrenchers decided one day to give Conrad and Bean a problem they were certain the men would be unable to solve. As the simulator showed the men within a minute of a lunar touchdown, the sim team fired their display with ERROR CODE 1202. As expected, Conrad and Bean and their Mission Control partners were unable to resolve the problem—error code 1202 was so obscure, no one had ever studied or trained for it. Conrad and Bean aborted the landing, and Dick Koos's simulation geeks chalked up a win for their side.

The 1202 alarm would never be experienced by the Apollo 12 crew or any other crew, with one exception. Armstrong and Aldrin would experience a 1202 (and a similar 1201) alarm several times during their descent. Though the sim team was proud of putting one over on the astronauts, they never thought to simulate a 1202 code with the Apollo 11 crew—the very ones who would soon find themselves in its crosshairs.

29

APOLLO 11

LAUNCH PAD 39A

It had taken four hundred thousand employees working for twelve thousand companies and corporations to reach a point where sending a manned rocket to the Moon could reasonably be accomplished. Of those twelve thousand contributing manufacturers, there were four prime contractors—one for each stage, plus the lunar module and the command service module. Boeing had been the contractor for the bottom stage, now fitted with five F-1 engines from Rocketdyne. The stage had been built at the NASA Michoud Assembly Facility in New Orleans. The second stage had been assembled by the newly merged North American Rockwell at its facility in Seal Beach, California. The third stage, also built in California, was assembled by McDonnell Douglas at its facility in Sacramento. North American Rockwell was also prime on the CSM, and Grumman was lead contractor on the LM.

Before NASA could assemble each of these five major sections of the Saturn V "stack," it first needed to get them to Cape Canaveral. The biggest piece—the first stage—made its way to the Gulf of Mexico via a barge along the Mississippi River. Once in the Gulf, the barge shipped the hardware around the Florida peninsula, up the eastern coast to Cape Canaveral, then negotiated a series of special canals to reach the Vehicle Assembly Building. The second stage had an even more convoluted journey. It was shipped from Southern California down to the Panama Canal, then northward through the Gulf of Mexico to Mississippi. Since North American had no way to do a full-up test of the stage and its engines in Southern California, the rocket was offloaded at NASA's John C. Stennis Test Center where the entire system and its engines could endure a full-up test. Once it was cleared for use and man-rated, it was then put back on the barge and sent around the tip of Florida to Cape Canaveral. The much smaller third stage had an easier and far quicker journey. As the Gemini program had begun to ramp up, a former U.S. Air Force pilot named John M. "Jack" Conroy foresaw the need for cargo planes that could haul bulky cargo. Together with aircraft salesman Lee Mansdorf, the pair created Aero Spacelines, a company whose sole purpose was to convert the Boeing 377 Stratocruiser into a flying cargo plane large enough to carry rockets for NASA. Located at the Van Nuys Airport in California, the two men found themselves in the right place at the right time, and Aero Spacelines began ferrying the first and second stages of the Gemini Titan rocket from the Martin Company's facility in Maryland to NASA's launch complex at Cape Canaveral. Due to its bulbous shape, the new plane was dubbed the Pregnant Guppy. Eventually the converted Boeing plane had to be bulked up even further to allow transport of the Saturn's S-IVB third stage, thus earning a new name: the Super Guppy.[1]

Despite the inherent risk in transporting such large and delicate cargo over such great distances, all the pieces of the Saturn V arrived successfully and without incident. Then the large team of technicians at the VAB began their work, painstakingly putting the complex machine together, piece by piece.

The Saturn V rocket that would launch the Apollo 11 mission into space and, hopefully, a lunar landing three days later, stood tall and silent in the pre-dawn darkness of July 16, 1969. The place from which it would leave Earth and begin its journey had been named Launch Pad 39A by NASA. The fact that launch pads at Cape Canaveral had become so numerous that one would be labeled "Launch Pad 39A" is a testament to how far the U.S. space program had come in only eleven years. There would eventually be more than forty launch pads (later renamed "launch complexes") built in the Cape Canaveral area; only a handful would still be in use a few decades later.

After going through the time-consuming chore of suiting up, Neil Armstrong, Mike Collins, and Buzz Aldrin followed Guenter Wendt into the launch gantry elevator. Guenter pressed a button, and the elevator began a steady climb toward the rocket's apex. Due to the cramped nature of the astronaut ingress procedure, Guenter dropped Aldrin off at a metal porch three-fourths of the way up, where the astronaut stood and watched as the elevator carried

Neil and Mike two floors higher. Buzz could hear the clang and bang of its arrival there—a place they called the "white room," where astronauts made their final preparations before being wedged into the Apollo capsule. He could hear the sounds of their boots on the gangway, the voices of the small crew as they assisted the astronauts into their flight seats. A few moments went by, then everything became very quiet.

It was still dark out, but the faintest hint of twilight was beginning to change the eastern sky, its horizon line as flat as the ocean that formed it. In a few hours that flat line would become rounded as they traveled high enough to see the curvature of the Earth. As he stood on the lower level, Buzz scanned from east to west—noting where hundreds of thousands of people had lined up on the beach to witness this historic launch. Hundreds of campfires lit by tourists who had partied there all night pierced the Florida morning air. And people. Some would later say a million people had shown up by the time the rocket lifted off. They could not see him, hidden as he was behind the many steel girders, but through gaps in the superstructure Buzz could see them. He contemplated the amazing lifelong sequence of events that had brought him here. He was one of three men out of billions that had been chosen—one of only two who would have the honor of the first Moon landing. Yes, it was an honor, but a terrifying one.

"THE MONSTER SPRINGS TO LIFE"

We are about to make history.

—Gene Kranz

Thy will be done.

—Wernher von Braun's prayer, moments before launch

Two weeks before the scheduled July 16, 1969, launch of Apollo 11, Wernher von Braun—needing a respite from the pressures that had been building for years—took his family on a cruise around the Greek islands,[2] making a stop in Delphi to visit the shrine of Apollo. There was more to the trip, however, than avoiding the pressures of work. Von Braun was disheartened over the loss of yet another good friend—noted space writer and fellow German expatriate Willy Ley, who had passed away three weeks earlier from a heart attack. Ley's science writing, especially in the area of human space travel, had done almost as much as von Braun to fire up the imaginations of Americans and the people of the world. And human flight to the Moon had been one of his more fanciful story subjects. Now Ley would not be around to see the culmination of his three decades of work.

The von Brauns returned to Huntsville, and on the July 13 Wernher and Maria flew to the Cape, staying at the Holiday Inn in Cocoa Beach. There followed three days of constant meetings with VIPs, government officials, visiting foreign dignitaries, and, of course, the media.

The night before the launch, Wernher was the keynote speaker at a Time-Life gala held at the Titusville Country Club. Due to gridlock resulting from a million people descending on Cape Canaveral to watch the launch, NASA was forced to fly von Braun to the event by helicopter. At a cocktail party that evening he met the playwright, and occasional space writer, Norman Mailer. As they discussed the future of the Apollo program, von Braun confessed his fear that the U.S. government would cancel the space program after the first manned lunar landing. Mailer told him, "Who are you kidding? You're going to get everything you want." Von Braun, however, was not convinced. Mercury had its next-generation follow-up in Gemini. Gemini had its next-generation follow-up with Apollo. But what was after Apollo? For the past year he had been pressuring NASA to come up with a post-Apollo plan that would make sense to the taxpayers and to Washington, telling the agency, "There's nothing in the program to make lunar visits meaningful."[3] Once again his idealism was being frustrated; there were no plans for permanent Moon bases or springboarding to Mars—all part of his lifelong passion.

That evening Wernher did not sleep well, and he eventually gave up trying. He arrived at the Launch Control Center at 4:00 a.m. It was July 16, and the launch was scheduled for 9:32 a.m. Wernher took his assigned place in the mission manager row, put on his headset, bowed his head, and prayed.[4]

In Texas, Gene Kranz woke up at 4:30 a.m. There had been no phone calls during the night, meaning all was going well at the Cape. As he showered and shaved, his wife Marta made him a large sack lunch. Dressed, alert, and ready, he kissed Marta good-bye and jumped in his car, heading for the MCC.

Upon arriving at Mission Control, Gene took the handoff from the previous Mission Control shift known as the Black Team. Its controllers had spent the previous twelve hours prepping the Saturn V for its historic launch. Though everything appeared normal, the atmosphere was both electric and tense. Chris Kraft's confidence in Gene's White Team was so high, the team was assigned to both the launch of Apollo 11 and the LM landing three days later—it was everything Gene had hoped for. All the dreams, all the planning, all the design, all the work, all the money, all the construction, all the testing, all the failures, all the successes—it all came down to this moment, this launch, and the next few days. Like von Braun, Gene Kranz quietly said a prayer.[5]

The countdown continued without any major hitches—about as flawless as even the sternest mission controller could expect. When Flight Director Cliff Charlesworth announced that just prior to launch at T-minus nine minutes he would be locking the MCC doors, a flood of controllers ran out for their last chance at a bathroom break.

In the command module, pilot Michael Collins was seated on the right side, with Neil Armstrong in the left seat and Buzz Aldrin in the center. In his book, *Carrying the Fire*, Collins described the launch in almost anticlimactic terms: "[M]y adrenaline pump is working fine as the monster springs to life. At nine seconds before liftoff, the five huge first-stage engines leisurely ignite, their thrust level is systematically raised to full power, and the hold-down clamps are released at T-zero. We are off!"[6]

The final countdown and launch of Apollo 11 were so perfect that in his book, *Failure Is Not an Option*, Gene Kranz dedicates only a single sentence to it: "The launch is flawless, as if this were just another simulation on a very good day."[7]

As the bottom stage depleted its fuel, Collins called, "Cutoff!" In Houston, the mission controllers now had to make their GO/NO-GO decision for orbit. Dave Reed, the flight dynamics officer for the launch, received quick input from the controllers in the trench. Everyone was unanimous—orbit was a GO. "Go, Flight. We are Go!" Reed announced.

The CSM and its three astronauts entered Earth orbit.

Immediately after the launch, Wernher von Braun joined Martha for two restful days meeting old friends. Then he flew off to Houston to be on hand for the landing.

TLI, TLC, AND LOI

Buzz seems to have gotten up on the wrong side of the bed this morning.

—Michael Collins's thoughts just prior to TLI burn

Whew!

—Neil Armstrong's reaction to the commencement of TLI

The next three major navigation hurdles to be overcome were Trans-Lunar Injection (TLI) in which the spacecraft would power its engine, break out of Earth orbit, and speed toward the Moon. Next would come Trans-Lunar Coast (TLC)—a period of unpowered flight where the astronauts simply coast to the Moon by inertia—defined so eloquently by Isaac Newton's first law of motion. Once they reached the Moon, the astronauts would then perform hurdle number three: Lunar Orbital Insertion (LOI)—firing their engine in the reverse direction to slow themselves down to a speed and place where the Moon would capture them in its orbit.

The schedule for the Apollo 11 mission was tight, and required the TLI burn only three hours after launch. As CSM pilot, this was Mike Collins's responsibility. The burn took place over the Pacific Ocean in a zone where Mission Control frequently experienced radio blackouts with its spacecraft. Not wanting to miss any part of the TLI burn, NASA had arranged for several jet aircraft to fly in circles high above the Pacific in order to forward Apollo 11's signal to Houston. Aldrin recalled that "three hours into the trip, Mike ignited the Saturn V third-stage rocket engines for our Trans-Lunar Injection burn to take us out of Earth's orbit. . . .The burn was successful."[8]

Once the five-minute, forty-five second TLI burn was complete, Apollo 11 was in Trans-Lunar Coast. Inertia would do most of the work for the next few days. Wasting no time, the astronauts switched seats, putting Collins in the left seat where the controls were mounted for separating the CSM from the Saturn V third stage, as well as the piloting controls for removing the LM from its third stage cavity.

During the LM docking and extraction procedure, the CSM experienced one unexpected glitch. As Collins attempted to swivel the CSM around and approach the Saturn V third stage, the computer refused to fully complete one of its programmed commands, forcing Collins to send and resend the command. He was finally able to line up the CSM and the LM, but expended 60 percent more fuel than had been budgeted for the maneuver.[9]

Once the LM was pulled from the third stage cavity, Collins piloted the two mated craft a safe distance from the upper stage, and Mission Control sent a command to steer the stage into a harmless solar orbit.

The astronauts breathed easier, doffed their pressure suits, and had lunch. It was 2:00 p.m., Cape time.

During the Apollo 8 mission several months before, Michael Collins's five-year-old son had asked him who was flying the spacecraft. Collins wasn't sure, so he had the CapCom send up the question to the astronauts. Bill Anders replied that "Isaac Newton was driving." Were it not for inertia and the laws of motion so aptly described by Newton, travel to the Moon would probably not be possible. As Apollo 11 coasted toward the Moon, pulling against Earth's ever-weakening gravity, they eventually reached a point where the Moon's gravitational pull was momentarily equal to that of Earth's—a point in space known as the equigravisphere. Once that point was crossed, it was "all downhill" to the Moon. Collins had been CapCom during the Apollo 8 mission when, for the first time in history, humans had been captured by lunar gravity. He had marked the moment by sending Commander Borman a simple message: "Welcome to the Moon's sphere."[10]

Now, as Apollo 11 passed the equigravisphere, the two mated ships began to speed up. From a low of 3,000 feet per second Apollo 11 would accelerate to almost 8,000 feet per second before the retro-burn of the SPS (Service Module Propulsion System) would begin to slow them down. The target speed for Lunar Orbital Insertion was 2,917 feet per second. In the CSM, all three astronauts prepared for the burn. LOI would actually include two burns—one to establish them into a wide elliptical orbit, and a second to alter their orbit to near circular. After the first burn completed, taking six long minutes, the astronauts checked their instruments to measure the accuracy of their orbit. To their amazement they had hit their target altitude and orbital speed with a margin of error of only one-tenth of one foot per second. Their subsequent conversation reflected their elation.

ARMSTRONG: That was a beautiful burn.

ALDRIN: Look at that: 169.6 by 60.9.

COLLINS: Beautiful, beautiful, beautiful, beautiful.[11]

A short time later, the astronauts initiated a second LOI burn, putting them in a circular orbit sixty miles high. As they pulled around the gray orb, the astronauts were able to get a view of the mysterious far side of the Moon, later described by Collins as "a jumble of tortured hills, cratered and re-cratered by 5 billion years of meteorite bombardment."[12] Then pulling back around to the Earth-facing side, they were able to examine their intended landing area on the Sea of Tranquility.

As the CSM coasted along its sixty-mile orbit, the astronauts followed the schedule plan, which called for eight hours of sleep. Then, after breakfast, Neil and Buzz would suit up and prepare for their lunar descent in the LM. So far everything was going better than anyone had expected.

ERROR CODE 1202

Neil Armstrong never raised his voice. He just saved his energy for when it was needed.

—Gene Kranz

Neil Armstrong is Czar of the ship.

—*Pravda*

The elevators at Mission Control had an odd propensity to occasionally get stuck between floors. It was a glitch that remained unrepaired as of July 20, 1969. When Gene Kranz arrived and entered the MCC building on the morning of the landing, he decided not to tempt fate, and took the stairs up to the control room. "This was not a day to get stuck in an elevator," he would later say.[13]

He passed well-wishers in the hallways. People gave him a pat on the back, or a quick "Good luck." Kranz was a thoughtful man who well understood the odd and unexpected series of events that had led him to this historic moment. "I have the same feeling every time I walk into the MCC. It is a place where history is being made, day by day. It is the home base, the control center of our explorers. As I continue down the hall, I get my usual vague feeling that somehow my entire life has been shaped by a power greater than me to bring me to this place, at this time."[14]

One item of lunar space history that has been underreported is that there was actually traffic in lunar orbit on the day Armstrong, Aldrin, and Collins began circling the Moon. Just before their arrival, a Soviet probe, Luna 15, had entered lunar orbit, preparing to land on the surface. It had been launched three days before the Saturn V/Apollo 11 launch from Cape Canaveral, presumably to steal some of America's lunar thunder. Fortunately, the Russians had cooperated with NASA and had given the agency their probe's planned trajectory so as to avoid a disastrous collision.

Armstrong and Aldrin, 240,000 miles away in the LM, had no time to concern themselves with lunar airspace traffic. Now fully suited up in their twenty-one-layer EVA pressure suits—the same outfits that would be used once they were walking on the Moon—they connected their oxygen hoses to the breathing source inside the LM and put on their helmets. For the next hour it would be the responsibility of Collins to go through a long checklist of activities that had to be accomplished before the two spacecraft could be separated. Armstrong and Aldrin could only wait.

In the CSM, Collins proceeded with closing the common hatches and disconnecting the electrical umbilical lines. Apollo 11 was about to divide into two separate spacecraft, which would then be officially referred to during all radio communications as the Columbia (the command module) and the Eagle (the lunar module). Alone in the CSM, he had no one to check his work—the mission's advancement now depended on the skills of a single individual. On his checklist was a requirement to set up a television camera at one of the windows to film the undocking sequence. Collins decided he was too busy, and scratched it off his list.

"There will be no television of the undocking . . . I'm busy with other things," he informed Mission Control.

"We concur," replied the CapCom.

There were several minutes of radio chatter between Collins and the LM crew, some of it jaunty. Finally, everything was ready. Apollo 11 coasted around to the far side of the Moon and out of radio contact with Mission Control. Gene Kranz and his team would not know the success of the undocking procedure until the astronauts came around again to Earth-side.[15]

As he readied himself to initiate the undocking, Collins threw his crewmates a jaunty instruction: "You cats take it easy on the lunar surface; if I hear you huffing and puffing, I'm going to start bitching at you."

"OK, Mike."

Collins vented the remaining air in the tunnel that connected the LM to the CSM. Then he threw the switch that controlled the undocking sequence, and the one ship became two. Despite Collins having vented the air in the tunnel, a small amount of residual air remained and gave a nudge to both spacecraft, separating them slightly faster than anticipated. Though small, this unexpected push would alter the LM's trajectory just enough that it would end up several miles off course.

As Collins watched through his window, the *Eagle* slowly backed away from its mother ship.[16] After they were a safe distance away, Armstrong briefly fired the steering thrusters to maneuver the LM into flight position. He then flipped a switch that fired a series of explosive bolts that released the LM's four spring-loaded landing legs. The mission now had a responsibility that only Collins could perform: a visual inspection of the LM. The mission could not proceed from this point without knowing that all four legs had extended and locked into their proper landing positions. After Armstrong rotated their craft so that Collins could get a 360-degree view, Collins gave his crewmates his report. "I think you've got a fine-looking flying machine there, *Eagle*, despite the fact you're upside down."[17]

Sixty miles above the lunar surface, the two ships coasted in formation at about four thousand miles per hour. Soon their orbit brought them around within view of Earth, and radio and telemetry with Mission Control were immediately reacquired. Their CapCom, Charlie Duke, was naturally curious as to how the undocking went.

"How does it look, Neil?"

Armstrong announced, "The *Eagle* has wings!"

It was not the first time a lunar module had flown independent and free from its mother ship—that had already happened with Apollo 10. But now began an historic moment—a descent and landing on the lunar surface. Gene Kranz went around the room, getting reports from each of his controllers. When everyone had reported "Go," he gave the green light to Duke for the next step: "You're Go for DOI."

DOI, for Descent Orbit Insertion, was a series of events that would cause the LM to slow its speed and allow gravity to pull it closer to the surface. The first maneuver in DOI was for Armstrong to tilt the LM so that its engine faced front—in the direction they were headed. When the engine fired, it would become a retro-rocket, decreasing their velocity. Like the undocking, trajectory and timing established by simple math required this maneuver to be performed while the LM was out of ground contact, around the far side of the Moon. At the appointed moment, the descent engine fired for precisely 28.5 seconds. Up to this point the onboard guidance computer had handled all the work of piloting the craft. The CSM was no longer visible—somewhere in orbit high above them. Armstrong and Aldrin's instruments indicated they had descended to an altitude of fifty thousand feet—the same altitude the Apollo 10 LM had settled at before returning to the CSM. At this point, the mission could be safely aborted—they could return to the CSM if needed. A GO/NO-GO decision now had to be made—should they attempt a landing? That step was referred to as PDI—Powered Descent Initiative—and would result in the LM once again firing its descent engine to break orbit, slow its velocity to near zero, and set down on the surface. On the ground Kranz once again polled his controllers. Again, each of them said "Go."

Due to changes in its attitude as the LM maneuvered into descent position, its antenna was not pointed directly at Earth, and so Charlie Duke relayed the decision to Eagle via Collins in Columbia. "You're go for powered descent," announced Duke.

Duke also recommended to Armstrong that he turn the LM slightly so that its antenna was in a better position—a suggestion which succeeded in greatly improving their communication and telemetry links. And as that telemetry started pouring in, Steve Bales, the mission guidance operator, began to notice something was off. The LM's speed was too fast, and it was not in the correct position.

"Flight," Bales reported. "We're out on our radial velocity; we're halfway to our abort limits. I don't know what's caused it, but I'm going to keep watching it." Further calculations showed the LM would land about three miles past its intended target—a rocky, cratered area that had not been well studied, nor planned for in the simulations. For a flight whose accuracy had been calculated to inches and milliseconds, it was a major error. Mission Control would later discover that it was the small amount of residual air in the connecting tunnel and the extra nudge it provided to the LM that threw off the ship's speed and trajectory.[18]

At 4:05 p.m. Eagle's computer fired up the descent engine, and the final stretch to the landing began. The atmosphere in the LM was professional, but tense—men were about to attempt to land on the Moon for the first time in human history. Carefully and steadily, Armstrong double-checked the computer's work as it maneuvered the LM closer to the surface. Aldrin continued to keep his eyes on the instruments, especially the altimeter, which steadily counted down: 40,000 feet . . . 38,000 feet . . . 36,000 feet. He continued to make occasional callouts of the readings.

With the LM's windows facing downward, Armstrong and Aldrin were able to make out a few landmarks—landmarks that did not match with the clock. That's when the men realized they were farther along than they were supposed to be.

"Our position checks downrange show us to be a little long," reported Armstrong.

Steve Bales told them what he already knew. "We confirm that."

The responsibilities between Armstrong and Aldrin had been established months before and were clear. Despite the fact that Aldrin's official title was lunar module pilot, it was Armstrong's responsibility as commander to pilot the LM starting from the moment it was disconnected from Columbia, all the way down to its lunar landing. Though Armstrong was the actual pilot, Aldrin had been trained to take over in an emergency. Aldrin's job was to keep his eyes fixed on the LM's gauges and readouts, verbally reporting to Armstrong all information he would need pertinent to the descent, such as attitude, velocity, and altitude. With Aldrin's callouts, Armstrong could keep his eyes off the gauges and focus on what was about to become one of the most difficult piloting jobs in history.

There was so much to think about, so much to watch out for, so many maneuvers to get right, so much training that had to be brought to bear. Then, just as the LM descended below the 34,000-foot mark, the quietude within the cabin was broken by the sound of the craft's master alarm. It was loud and brash and unmistakable. The last thing these two men needed at that crucial moment was a distraction, and now a giant one was suddenly thrust upon them. The LM's systems had detected a major problem.

The LM's computer was so ancient by today's standards that information was outputted in four-digit codes rather than English sentences. The control panel now displayed one of those four-digit codes: 1202.[19]

"Twelve oh two," said Armstrong, reading the number on the display.

"Twelve oh two," repeated Aldrin.

The problem these men faced was exacerbated by another unforeseen dilemma: neither of them had ever experienced this error code during their extensive, exhaustive training regimen. What the hell was a 1202? Neither of them had a clue.

At Mission Control, no one had to be woken up. Everyone—even those not scheduled to work the shift—had assembled together to watch this historic moment. All the NASA brass and managers, and a slew of astronauts, including John Glenn, were there. It was standing room only, and now some drama filled the room: a last-minute alarm and a code number that no one on the ground recognized were demanding immediate attention and resolution. The reputation and image of the unruffled, no-panic NASA mission controller was being challenged like nothing that had ever come before. Everyone looked to their neighbors for a solution, but all they received was the same puzzled expression, as if to say, "Now what?"

Steve Bales was born in a small town in Iowa, the son of a janitor father and a beautician mother. At the age of thirteen he began tuning in to watch *The Wonderful World of Disney*, where he was introduced to Wernher von Braun and his starry predictions of manned space travel. After a few episodes, Steve knew what he wanted to do with his life. After high school he enrolled in the aeronautical engineering program at Iowa State University, where he graduated four years later with a bachelor's degree. He was hired by NASA in December 1964.

NASA assigned him to be a flight controller in charge of flight dynamics. It was his job to always know the precise location of a spacecraft in space—a crucial responsibility. Monitoring the guidance systems of those craft was a major part of his job. He served as a backup controller for two of the early Gemini missions, then got his turn at the desk for Gemini 10. He continued to work as a flight controller during the Apollo program.

Apollo 11 would put him in the history books.

Like many flight controllers, Bales had GO/NO-GO authority in certain situations. For the Apollo 11 mission he was assigned to be the guidance operator, what was then whimsically referred to by the acronym GUIDO. As the GUIDO in charge during the landing, he was one of the most important mission controllers in the building. While monitoring the LM's descent he noticed it was traveling about twenty feet per second too fast—a potentially fatal problem if not corrected. As he pondered what to do about that problem (GO or NO-GO), the 1202 alarm appeared on his screen and was relayed to him verbally through his headset.

Over the intercom came the words of Neil Armstrong who was, as was his custom, cool as air conditioning. His request was direct and businesslike. "Give us a reading on the 1202 program alarm." Everyone turned to look at Steve Bales.

The problem was, even Bales did not know what a "1202 code" meant. Or at least he could not remember.

Standing at his workstation, Gene Kranz waited for his team to advise him. Kranz would later write of this moment that the alarm "had the capacity to create doubt and distraction, two of a pilot's deadliest enemies."[20]

Within Bales's support team was a twenty-four-year-old computer specialist named Jack Garman. A University of Michigan graduate from Illinois, Garman had been hired by NASA right out of college. He found his calling in onboard computers and ended up working with MIT in the design and testing of the Apollo Guidance Computer. During the period of simulation testing of the onboard computer systems, Gene Kranz had instructed Garman to make a list of every possible error code, and the proper response to each one. Garman then placed that list under a pane of plexiglass atop his desk. When the word came down that the onboard computer was issuing a 1202 error code, Garman knew right away what it was, and what to do about it. A 1202 error code, he knew, meant that the LM's computer was overloaded with work—more computing demands than it had capacity for. Garman already had the answer ready: GO. There would be no reason to abort as long as the 1202 error coded was not continuous. He relayed this to Bales, who relayed it to Kranz, who relayed it to CapCom Charlie Duke, who relayed it to the two astronauts. It is testament to the amazing management procedures set up by Kraft and Kranz that the entire 1202 episode took less than twenty seconds from discovery to resolution.

Over the next few minutes, as the LM continued its descent from the alarm-triggering 33,500-foot mark to a lunar-kissing 3,000 feet, several more 1202 alarms sounded, and in each case Garman and Bales insisted the landing was still GO. Aldrin and Armstrong were close enough now they could visually examine the terrain for a satisfactory place to land.

Then a new alarm sounded.

"Program alarm!" said Aldrin. "1201." With the Moon coming up fast and the adrenaline pumping faster, the coolness in their voices was beginning to slide. In both the LM and at Mission Control there was frustration with the alarms and the dangerous distraction they were posing. When Garman heard the "1201" he let Bales know immediately that 1201 was the "Same type!" The landing was still GO.

A crucial part of the Apollo 10 mission had been to photograph the landing approach for Apollo 11. As part of their training, Armstrong and Aldrin had pored over those photographs, memorizing all the significant rills, hills, craters, and boulders. These images would become the landmarks they would watch for as they approached their landing. Yet so much time and attention had been taken up addressing the 1202 and 1201 alarms that both men had been too busy to watch for those surface markers.[21] Once they realized that the alarms were not serious and they could ignore them, most of the important landmarks had already passed beneath them.

At the moment the CSM had detached itself from the LM, the LM's propellant tanks contained exactly twelve minutes' worth of fuel and oxidizer—theoretically enough to descend to the lunar surface with about ninety seconds of fuel to spare. The extra propellant was intentional as it was assumed there might be a slight delay while the astronauts scouted for a safe, boulderless, craterless landing area. In hundreds of landing simulations, many of which included landing delays for such maneuvering, they had always landed the LM by now. Yet here they were, seven hundred feet above the surface, and still flying. The 1202 alarms had been a time-eating distraction, and the alarms were sounding more frequently as the already overloaded computer memory had even more demands placed upon it. Unexpectedly, the LM was running out of fuel.

Robert Carlton was a mission controller who graduated from Auburn University with a degree in mechanical engineering. Like so many others in NASA, he had been recruited right out of college. He was thrust into the Gemini program where he was forced to learn by doing. So busy had the Gemini mission controllers been that they had no time to study the designs for the upcoming Apollo program. According to Carlton, back then all the controllers had to go on was "a picture on the wall" of what the Saturn V would look like.[22]

By the time Apollo 11 came along, Carlton had worked his way up to an important desk in the flight control room, a position referred to as LM CONTROL. It was his responsibility to monitor fuel and oxidizer consumption in the lunar module. Now, with the astronauts descending past the one-hundred-foot altitude level, Carlton's control panel readouts gave a warning he wasn't ever supposed to see, which forced him to make an announcement he was never supposed to give. "Sixty seconds!" he reported into his headset, but loud enough so everyone could have heard him without the mic.

To this point the LM had used far more descent propellant than anticipated. In the simulators back on Earth the astronauts had trained for what would happen if the LM ran out of fuel, and the simulations all involved aborting the mission. Now, as Carlton stared at his data, he knew there was a real possibility the world's first manned lunar landing might fail.[23] Carlton's announcement was relayed to Charlie Duke, who relayed it to the astronauts. Of this moment Gene Kranz would later write, "I never dreamed we would still be flying this close to empty."[24]

A later analysis would show that the cause of the 1202 overload on the LM computer was an out-of-procedure decision made by Aldrin to leave the rendezvous radar on during their descent—a decision that made him feel better at the time, but gobbled up enough extra data to overflow the computer.

LUNA FIRMA

> *Every mission had a series of little traumas.*
>
> —Robert Carlton, Mission Control engineer

> *I changed my mind a couple of times again, looking for a parking place.*
>
> —Neil Armstrong

"Pretty rocky area," said Armstrong, scanning the surface ahead.

"Six hundred, down at nineteen."

There is an old pilot's axiom: when in doubt, land long. And that was what they were about to do. NASA had programmed the LM's guidance computer to set down in a predetermined place that lunar probes and Tom Stafford's photographs from Apollo 10 had seemed to indicate would be ideal. But now that they were just a few hundred feet from the surface, all sorts of unforeseen surface boulders and craters came into view. Aldrin looked up briefly to see what Armstrong was seeing. He realized their guidance computer was pointing their landing toward an area too risky for a safe touchdown. He later described what happened next.

> Neil had made up his mind. "I'm going to . . ." He didn't have to finish his statement. I knew that Neil was taking over manual control of the Eagle. Good thing, too, since our computer was leading us into a landing field littered with large boulders surrounding a forty-foot-wide crater. Neil made a split-second decision to fly long, to go farther than we had planned, to search for a safe landing area.[25]

"Okay, four hundred feet, down at nine." Aldrin decided it was time to start calling out forward speed as well. "Fifty-eight forward."[26]

"No problem," said Armstrong.

But Aldrin could sense there was a problem—several of them, in fact. He could tell his pilot was struggling to find a safe place to set down their craft, and their propellant tanks were approaching empty. As if reading his mind, Armstrong asked for a report.

"Okay, how's the fuel?"

"Eight percent." Almost gone.

"Okay, here's a . . . looks like a good area here."

Glancing out his window, Aldrin could now see the LM's shadow on the lunar surface—they were close.

Then another problem.

"Two hundred fifty. Altitude-velocity lights." In LM jargon, this meant that the warning lights had come on indicating they were not getting accurate radar data. The entire landing was becoming more seat-of-the-pants every moment.

Then Armstrong spotted a flat, boulderless area to shoot for. "Finally we found an area ringed on one side by fairly good-sized craters, and on the other side by a boulder field. It was not a particularly big area, only a couple of hundred square feet, about the size of a big house lot."[27]

"Gonna be right over that crater."

"Two hundred feet, four and a half down."

"I've got a good spot."

But just because Armstrong had finally found a suitable landing area, did not mean they would have enough fuel left to get there. They were rapidly approaching what the astronauts referred to as the "bingo fuel call." This was the point where the propellants had become so depleted that the mission would have to be aborted, the ascent engine fired, and the LM forced to return to Mike Collins and the CSM.[28] Aldrin did not bother to remind his pilot of this—he knew that

Armstrong was all too aware. Instead he used another method: "Without wanting to say anything to Neil that might disrupt his focus, I pretty much used my body English, as best I could in a spacesuit, as if to say, *Neil, get this on the ground!*"[29]

Back home in Mission Control, CapCom Charlie Duke was watching his instruments, which indicated the same measurements Aldrin was seeing. Duke realized time was running out, and he sent up a calm reminder of their remaining fuel, measured in time.

"One hundred, three and a half down, nine forward," came the reply.

The LM was slowing its descent rate in preparation for touchdown, and at the seventy-five-foot level Kranz mentally worked out the math. The astronauts were descending at a rate of two and a half feet per second, which would require at least thirty seconds of fuel. Kranz ordered all chatter to cease except for Carlton's fuel callouts, and as everyone in Mission Control went silent as a tomb, the radio transmissions from the LM went silent as well—the astronauts were too busy to talk. Then a single comment from Aldrin. "Forty feet . . . picking up some dust . . . thirty feet . . . seeing a shadow."

As the clock ticked relentlessly onward, Carlton announced, "Thirty seconds," and Duke relayed the information to the crew.

Normally the astronauts would have appreciated such updates, but it was crunch time and the updates were just a distraction to the two highly focused men. Aldrin took another quick glance out his window and saw the shadow of one of the three footpads—and it was touching the surface.

From Carlton: "Fifteen sec . . ."

Aldrin then made the announcement everyone had been waiting for: "Contact light." He later recalled, "Neil and I looked at each other with a stolen glance of relief and immense satisfaction. The LM settled gently, and we stopped moving. After flying for more than four days, it was a strange sensation to be suddenly stationary."[30]

"Shutdown," said Armstrong, announcing the shutdown of the descent engine.

Aldrin verified. "Okay, engine stopped."

Ignoring the emotions that would have consumed most people, Armstrong and Aldrin went right to task, working on their postflight checklist.

"ACA out of detent," said Aldrin. This was a reminder to his pilot to take the Attitude Control Assembly—the joystick Armstrong had used to pilot them to the surface—out of MANUAL control mode and return it to AUTO mode, in preparation for their later takeoff.

"Out of detent. Auto," replied Armstrong.

As the men continued with their checklist, everyone back on Earth was wondering what was going on. It seemed that the Eagle had landed safely, but where were the historical pronouncements, the cheering, the football-stadium hoopla? Carlton reported to his superiors what his console was telling him: "Flight, we've had shutdown."[31] Either a landing or a major malfunction. But which?

In the LM, Charlie Duke's voice came over the radio, reminding the astronauts that Mission Control and everyone back home were awaiting word.

"We copy you down, Eagle." As much a question as a statement.

Armstrong paused from their checklist work to reply.

"Houston, Tranquility Base here. The Eagle has landed."[32]

And *now* came the football-stadium hoopla as the Mission Control engineers stood, applauded, and cheered—so much work and effort, finally at fruition. Duke advised Armstrong and Aldrin of how tortuous the suspense and anticipation had been for them.

"You got a bunch of guys about to turn blue. We're breathing again."[33]

Gene Kranz would like to have joined the cheering, but he knew there was still much that required immediate attention, including the STAY/NO-STAY decision. For the moment, he was too choked up to speak or think.

Neil's voice came through with a plea for quiet and time: "I tell you; we're going to be busy for a minute."[34] "Busy" because they had a series of tasks and decisions to make that were simply frantic. If there was some malfunction of the LM that would force them to leave, they had two very tight windows of opportunity to make an emergency takeoff and bring their vessel within rendezvous range of the CSM. Those two points were three minutes after touchdown, and twelve minutes after, so there were a great many instrument and status checks that had to be made within the first three minutes. A NO-STAY decision could be made by either Mission Control or the LM's crew. The STAY decision would be determined by Mission Control. With the three-minute window approaching and all systems on the ground reading GO, Charlie Duke radioed Eagle their first decision.

"You are STAY." They were good to remain on the lunar surface for at least twelve minutes.

In Boston's Fenway Park a major league baseball game was underway. The umpires suddenly signaled a time out, and the announcement was made that the astronauts had landed safely on the Moon. The audience broke into wild applause and cheers. In summer camps throughout the country, young boys and girls had watched the landing on small television sets. In department stores, employees and customers alike had flooded to the electronics section to watch the landing on multiple TVs. In New York and Las Vegas—cities that never sleep—everything came to a halt as people tuned in. Not everyone was so enthralled, however. At a bar in New York City, a bartender had been excoriated by two customers when he attempted to change the television channel to the Apollo 11 Moon landing, one of them shouting, "We don't care about no Moon!"[35]

Michael Collins could now claim the title of Loneliest Man in the Universe. As the sole occupant of a spacecraft orbiting a foreign body a quarter-million miles from Earth, he set a new standard for what it meant to be alone. In a pre-flight press briefing a reporter had commented, "Not since Adam has any man experienced such loneliness." And yet Collins did not feel alone. "Far from feeling lonely or abandoned," he would later write, "I feel very much a part of what is taking place on the lunar surface. . . . I feel this powerfully—not as fear or loneliness—but as awareness, anticipation, satisfaction, confidence, almost exultation. I like the feeling."[36]

Yet after the flight he would make a telling comment to his fellow crewmember, Buzz Aldrin: "I knew I was alone in a way that no earthling had ever been before."[37]

An hour earlier, Collins had watched his two colleagues pull away from the command service module and begin their descent to the lunar surface. Collins had been able to watch as the Eagle fired its descent engine to reduce velocity, then passed beneath Columbia. After exchanging a few words with the CapCom in Houston, he initiated a period of radio silence, knowing that from then on the only radio chatter that really mattered was between the ground and the LM. No longer responsible for the LM and its crew, he spent his time checking and rechecking the condition of his craft, happily discovering that it seemed to be functioning perfectly. Using a sextant, he had checked his position relative to a well-known lunar landmark, and the position of the LM when it was nothing but a tiny dot far below, still on its landing approach. He had continued to monitor the communications between Houston and the LM, and therefore was aware of the 1202 and 1201 alarms as they occurred. Like the LM crew, he had no idea what the alarm codes referred to. He had listened intently to the reports from the LM, monitoring Aldrin as he called out the altitude and velocity values. He had heard Aldrin announce, "Contact light" and, like everyone else, was puzzled that the first thing they did next was proceed with the postlanding checklist. Had they landed safely or not? Then Armstrong's voice came in reassuringly.

"Houston, Tranquility Base, here. The Eagle has landed."

Collins exchanged a few words with Houston, then once again left the frequency open for the LM crew.

Soon the Columbia circled behind the Moon, cutting Collins off from any contact with his colleagues, with Earth, or any human being anywhere. Below him the Moon moved into night—creating a dark black void, like a round hole in the universe. On his next orbit Collins planned to spot the LM with his sextant in order to record its exact location—a location different than what had been planned. The new landing location would slightly affect the takeoff and rendezvous to come, as the LM's starting point was now altered from the original flight plan. All this would need to be plugged into the ship's computer, and soon.

The next orbit would take two hours, but Collins had plenty to do in the interim.[38]

On Earth, the television audiences around the world were left out of what was then, and still is today, a very misunderstood part of the lunar mission: the extensive preparation required to open the LM hatch, descend, and explore. Owing to TV editing, many people believe today that Armstrong and Aldrin left the LM to do their lunar walks right after landing. In reality, so much meticulous preparation was required that a full seven hours elapsed from the time the LM touched down until the moment Neil Armstrong exited the craft. When the astronauts finally did descend to the lunar surface, six hundred million people back home—one-fifth of all humanity—would be watching.

As long as there were no problems in Australia.

The transmission antenna on the Eagle was small, and its radio/TV transmitter was not high in power. Therefore, NASA needed an extremely large and sensitive antenna on Earth to receive the astronauts' weak radio and television signals. There were three radio dish antennas that would not only be capable of receiving those signals, but would also be facing the Moon at the right moment. Two of them were in Australia, and the third was the Goldstone array in California. The one that NASA was putting its money on—the one that had the ability to forward them the best quality signal—was the Parkes Radio Observatory in southeastern Australia.

Neil Mason was the Parkes Observatory staff member whose responsibility at the dish control desk was to keep the dish aimed in exactly the right direction. Since both the Earth and the Moon were in constant motion, this was no easy task, even under ideal circumstances. As the world's first Moon walk was about to begin, a huge windstorm blew across New South Wales, tugging at the dish and threatening to topple it over. Normally the dish supervisors would have rotated the dish into a safe position in such a windstorm, but this was the Moon landing, and they had a responsibility to history they intended to fulfill. NASA used the Parkes signal for the entire two and a half hours of the Armstrong/Aldrin Moon walk. What few people would ever know was that during that entire time, the Parkes dish operated well outside its prescribed safety limits. As a favor to their friends in America, Australian scientists took a big risk for the sake of recording world history.[39]

So focused was he on the job of aiming and holding the dish in place during the windstorm that its operator, Neil Mason, never once glanced at the observatory's television monitor—he missed the entire Moon walk so that everyone else on planet Earth could see it.[40]

Two hours had passed since Mike Collins had said good-bye to his two crewmates. Coasting in orbit in the CSM, Collins was coming around from the far side of the Moon and approaching a point where Mission Control said he should be able to spot the LM. The engineers at NASA were puzzled over where the LM had landed, and they needed Collins to help them out. They knew the ship had landed somewhere beyond its original target area, but how far beyond? It was crucial to know where the LM was located because once it took off from the Moon it would need to rendezvous with the CSM, and to program the flight and trajectory of that rendezvous into the ship's computer, the engineers needed to know the LM's starting point. As Collins passed over the suggested zone, he used every skill and tool he had to find the LM on the surface, but to no avail. Collins would have three more chances to spot the LM before his colleagues stepped onto the lunar surface.

The time arrived to depressurize the LM's cabin and walk outside. Before exiting, it was important that there be no pressure differential between the inside and outside. This meant that all the oxygen in the LM had to be evacuated; like the environment outside the LM, the interior had to be a vacuum. This was due to a safety feature built into the LM's design whereby the exit hatch could only be opened inward. As long as there was air pressure inside the cabin, that pressure would always push against the door, keeping it sealed tight. Thanks to the Law of Unintended Consequences, this safety feature now posed an unforeseen dilemma. As the two astronauts vented the oxygen from their vessel, the pressure gauge creeped slowly downward until it reached 0.10 psi—almost, but not quite, a vacuum.[41] They waited, but the remaining oxygen within the LM refused to vent, and the pressure stayed at 0.10. Aldrin recalled,

> I watched carefully as the [internal oxygen] gauge eased down toward zero. I attempted to . . . open the hatch, but it wouldn't release. The cabin still wasn't quite empty of oxygen. Amazingly, just a tiny bit of oxygen pressure would keep that hatch from opening inward.[42]

Aldrin and Armstrong made several more attempts to pull open the hatch, but the tiny residual oxygen gas was enough to keep it in the closed position.

Had they come all this way only to be kept from leaving the LM and walking on the Moon?

Armstrong would later write about the hatch experience, chalking it up to the problem of untestability—one of the few things in the mission they were never able to rehearse before the flight was putting a LM in a vacuum environment to test the egress procedure: "That's a part of the exercise we never got to duplicate in any tests or simulations. . . . We had trained for this aspect of the mission in a chamber and under somewhat different conditions."[43]

Both men would later theorize where the persistent 0.10 psi was coming from: their own pressure suits. Their suits were designed to exhaust waste gas into the LM environment, causing them to replenish the gas that would bleed off. Eventually they were able to pull the hatch open, but the experience made Aldrin realize there could be another unforeseen problem. What if they closed the hatch and could not reopen it later to reenter the LM? In other words, if they weren't careful, a scenario whereby the astronauts might be stranded forever outside the LM, unable to ever re-open the hatch, get back inside, and fly home, was a plausible outcome. Aldrin recalled, "I made a mental note . . . if there was any oxygen remaining in the *Eagle* when I stepped out, and the hatch should close . . . the pressure inside would seal the hatch."[44]

After seven long hours of preparation, the men were ready to leave the LM and walk on the Moon. Armstrong got on his hands and knees, backing slowly out onto a flat "porch" just outside the hatch. Aldrin helped steer him down, guiding the bulky suit and its backpack so nothing snagged on the door or anything else. Armstrong then descended the ladder, carefully and methodically. Part way down he pulled a lever that opened an equipment bay door and triggered a television camera.

In Mission Control one of the newer astronaut recruits, Bruce McCandless, was at the CapCom desk waiting, like everyone else, for the TV camera's signal. When the image of Armstrong on the ladder appeared, he sent word up to the crew: "We're getting picture on the TV."

A few moments later, Armstrong reached the bottom of the ladder, whose step hovered three feet from the lunar surface. This required Armstrong to make a small jump to reach the next goal—the LM footpad. From there it was a simple step to the surface of the Moon. He then uttered the well-known words, "That's one small step for man . . . one giant leap for mankind."

Armstrong took a look around him, then gazed down at his boots. "The surface is fine and powdery," he reported. "I can kick it up loosely with one toe."

Armstrong informed Mission Control, and most of planet Earth, that his boot only sank into the surface dust about one-eighth of an inch—finally answering the long-debated question about lunar soil consistency. He then proceeded to collect a few rock samples. In the event they were forced to leave unexpectedly due to some emergency, they did not want to return home empty-handed. Once the initial samples were collected, Aldrin used a cable and pulley system to lower a 70-millimeter Hasselblad camera down to Armstrong.

After Armstrong had been on the surface for about twenty minutes, it was Aldrin's turn to descend. Like Armstrong, he was able to climb down without difficulty. Before going too far, however, he partially closed the hatch. Remembering their experience with the residual oxygen, he made sure not to latch it.[45] Then Aldrin continued his descent.[46]

Soon after reaching the bottom, a bodily need impeded his thoughts. "I decided this would be an excellent opportunity to relieve the nervousness in my bladder. I don't know that history grants any reward for such actions, but that dubious distinction is my 'first' on the Moon."[47]

Then Aldrin turned around to take a look for himself at the broad lunar landscape.

> In every direction I could see detailed characteristics of the gray ash-colored lunar scenery, pocked with thousands of little craters and with every variety and shape of rock. I saw the horizon curving a mile and a half away. With no atmosphere there was no haze on the Moon. It was crystal clear.[48]

Once the brief sightseeing moments were done, the two men set to work. Aldrin took the Hasselblad camera from Armstrong and shot a number of photos of the LM's landing struts and the ground around them. These photos were required by NASA for making landing decisions on upcoming flights. Then he returned the camera to Armstrong, who was the mission's designated photographer. A few more photos, then the astronauts set up the TV camera about fifty feet from the LM in order to record and transmit better views of "Tranquility Base" and its two lone occupants.

Then there were some ceremonial duties. Back on Earth a plaque had been attached to one of the LM's legs. Optimistically dated July 20, 1969, it commemorated the first lunar landing with these words: HERE MEN FROM THE PLANET EARTH FIRST SET FOOT ON THE MOON. WE CAME IN PEACE FOR ALL MANKIND.

Then they planted an American flag—its stars and stripes held aloft by a telescoping arm. Pushing its pole into the lunar soil proved a difficult job, but with effort they were able to get it to stand firmly. After taking photos of themselves posing with the flag, they tried walking, hopping, and even running, to see what human locomotion was like in the one-sixth gravity environment. Their movement experiments were interrupted by CapCom McCandless, who radioed in with an unexpected message: "Neil and Buzz, the President of the United States is in his office now and would like to say a few words to you."

When the message that Richard Nixon was calling came through his helmet's radio, Aldrin stopped cold. It was unexpected—it had never occurred to him that they would get any sort of civilian phone call while on the Moon, let alone one from the president. As he contemplated what to say next, Armstrong's voice came through: "That would be an honor."

The two astronauts positioned themselves in front of the TV camera, and the president spoke. "Hello, Neil and Buzz. I'm talking to you by phone from the Oval room at the White House."[49]

Armstrong and Aldrin patiently listened as the president spoke a few words of encouragement and congratulation, ending with, " . . . and all of us look forward to seeing you on the *Hornet* on Thursday."

Aldrin signed off with, "I look forward to that very much, sir." The astronauts saluted to the camera, then returned to their assigned duties.

Months before, Mission Control had meticulously planned out the first lunar exploration schedule, and it called for a mere two and a half hours outside the LM. After the brief conversation with their Commander in Chief, Aldrin set up three scientific experiments. For the sake of safety, the equipment needed for these experiments was compact and lightweight. Until the flight viability of the LM was real-world tested by Armstrong and Aldrin, NASA wanted to keep the ship's cargo weight down to a minimum. The three experiments consisted of a passive seismometer, a solar wind detector, and a laser reflector known as the Lunar Laser Ranging Reflector. The purpose of the LLRR was to determine the distance between the Earth and Moon to a very high degree of precision. Transmitters on Earth would fire a laser beam at the device and calculate the time it took for the beam to be reflected back to Earth. Using the reflector, NASA, JPL, and other science organizations and universities would later calculate the average Earth-Moon distance at 238,897 miles (384,467 kilometers). Amazingly, the LLRR would still be performing its function decades later.[50]

The purpose of the seismometer was to measure the number and size of meteoroid impacts, the extent of underground volcanic activity, and the interior makeup of the Moon itself. It was data from the Apollo 11 seismometer that would later show the Moon's structure was similar to Earth's with a crust, mantle, and fluid core.[51] The seismometer was not destined to enjoy the same longevity, however, as the LLRR. It operated for only three weeks before shutting down. Even so, during its brief lifetime the Apollo 11 seismometer revealed many of the Moon's geologic secrets. More sophisticated seismometers would be set up by future Apollo astronauts.

The solar wind experiment was designed to detect electrically charged particles that are naturally transmitted by the Sun far into space, but which are mostly deflected by Earth's magnetic field. The very simple experiment set up by Aldrin was a sheet of aluminum foil, one foot wide and four and a half feet long. It was mounted on a pole facing the Sun where it would collect, like a net, a number of solar particles. Before leaving the Moon, Aldrin would roll up the foil and pack it away for a return to Earth where scientists would study its results.

As Aldrin went about the work of setting up the three experiments, Armstrong was busy taking photographs[52] and collecting geologic samples. It was difficult to do both, so after Aldrin had completed his setup tasks, Armstrong handed him the Hasselblad and focused on rock collecting. He and Aldrin had begun their astronaut training in geology in March 1964—more than five years earlier. The training had been performed by career geologists as well as one fellow astronaut: Jack Schmitt, who would one day walk on the Moon as a crewmember of Apollo 17.

Eventually Aldrin took over the camera, and decided it would be a good idea for posterity to take a photo of a single footprint in the lunar soil. That photo, so definitive and human, became one of the most famous photos in history. Not to be outdone, Armstrong would also end up with a similar historic photograph. It shows Aldrin standing on the lunar surface with the LM, Armstrong, Aldrin's shadow, and their set-up experiments all reflected in Aldrin's visor.

All too soon, the two explorers received the three-minute warning from Houston: it was time to ascend the LM ladder and prepare to rejoin Collins in orbit. They re-entered the LM in reverse order of their exit—Aldrin first, and then Armstrong. But as Aldrin began to ascend the ladder, Armstrong reminded him of a small ceremony they had planned.

Edwin "Buzz" Aldrin had become an astronaut for only one reason—his friend Ed White had suggested it. Were it not for White, Aldrin would never have become an astronaut, and would never have walked on the Moon. In his pocket was a pouch containing a patch from the Apollo 1 mission, two medals commemorating Soviet cosmonauts Yuri Gagarin and Vladimir Komarov, and a small silicon disk with goodwill messages from seventy-three nations, including the Soviet Union. In addition, there was a gold pin in the shape of the olive branch of peace. Aldrin removed the pouch and tossed it onto the lunar surface.

After re-entering the LM, Aldrin used a pulley system to assist Armstrong in bringing up the boxes of lunar rocks. Then, Armstrong rejoined Aldrin in the LM.

At liftoff, every pound of weight that could be saved was an advantage, so after pressurizing the cabin, the astronauts removed their heavy life-support backpacks. Then they plugged their suits into *Eagle's* oxygen tank, depressurized the cabin again, tossed the backpacks and some miscellaneous trash out the hatch, resealed the LM, and repressurized.

Then it was time for dinner—cocktail sausages and fruit punch—after which Mission Control had scheduled in some sleep time. As Aldrin curled up on the floor and Armstrong leaned back on the ascent engine cover, Aldrin noticed a circuit breaker lying on the floor. A quick check of their instruments showed it was a crucial breaker needed to fire their ascent engine. Somehow it had been forced out of the control panel. They notified Houston, who told the astronauts to not yet attempt a repair—Mission Control would work out a solution. The astronauts were ordered to commence their sleep schedule. But with gravity back in play, a cold cabin, a cramped environment with no place to stretch out or lie down, and now the stress over a major component failure, sleep was close to impossible. Aldrin would later describe this sleep period as "a fitful state of drowsing."[53]

In the CSM above, Mike Collins had no trouble sleeping at all.

A few hours later, Houston radioed in the crew's wake-up call. Aldrin and Armstrong were not happy to discover that Mission Control had not yet come up with a fix for the recalcitrant breaker switch. Aldrin decided not to wait. Using a felt marker from their gear kit, he managed to push the breaker back into its cavity. Problem solved.[54]

Then it was time for Aldrin and his CapCom, Ron Evans, to proceed through a lengthy prelaunch checklist. On that checklist was a new item: turning off the rendezvous radar. The astronauts had not yet been informed of it, but Mission Control had figured out that it had been Aldrin's insistence on leaving the rendezvous radar on during their descent that had overloaded the LM computer and caused the 1202/1201 alarms. Aldrin did not like turning off the radar, but he acquiesced.[55]

As Armstrong and Aldrin were making preparations to fire their ascent engine to return to lunar orbit, the Soviet Luna 15 probe, circling the Moon high above them, fired its braking rocket and began its descent for an intended soft landing. No one knows for certain, but there was speculation that Luna 15's landing radar malfunctioned, failing to steer it away from a mountain. Whatever the cause of the malfunction, the probe crashed into the Mare Crisium, a mere two hours before Armstrong and Aldrin were scheduled to take off from the Moon to rejoin Collins in orbit. Had the Eagle landed about five hundred miles to the east, the astronauts might have scored another first—the first humans to witness a man-made object crash-land on the Moon.

The juxtaposition of the spectacular success of the manned Apollo 11 mission against the dismal failure of the Soviet Luna 15 probe put a period at the end of the Space Race sentence. The race—unofficial as it was—was over, and there was no doubt who had won.

All that was left was to bring Armstrong and Aldrin home safe.

HOME

> *The magnificent desolation of the Moon was no longer a stranger to mankind.*
>
> —Buzz Aldrin

> *Pilots take no joy in walking. Pilots like to fly.*
>
> —Neil Armstrong

Sixty miles above the surface, Collins continued to orbit the Moon. Every time Collins passed over the Mare Tranquillitatis, he had attempted to use scopes, instruments, and the naked eye to spot the lunar lander, but he was never able to do so. He had no trouble listening in on their radio chatter, but it would have been nice to have gotten a visual, and maybe even an aerial photograph. Even so, Collins would return home with literally thousands of photos of their target orb.

Minutes earlier, Collins had been awoken by Ron Evans, who informed the lonesome astronaut that they had managed to locate the Eagle—it had landed four miles past its target landing site. The good news was that they could calculate both ships' trajectories for rendezvous, and Evans wasted no time in reminding Collins they had a great deal of work to do in preparation for that maneuver. On his list were no less than 850 computer keystrokes to enter—and every one had to be perfect. As he worked through his checklist, Collins heard Aldrin's launch countdown reach "ten . . . nine . . . eight . . ." Collins paused to listen—this was a rocket engine burn that absolutely had to work. Aldrin's count reached zero, then he announced that the ascent engine had successfully fired—Aldrin and Armstrong were on their way to orbit.

Collins returned to his checklist.

Just before firing their ascent engine, both Neil and Buzz took one last look at the repaired circuit breaker. Would their makeshift unauthorized repair work?

The question was soon answered as the count reached zero and the upper stage of the lunar lander rose swiftly upward. Below them the lunar ground moved away at high speed. Armstrong and Aldrin completed two orbits before spotting Columbia. Their approach was careful and methodical, taking just under four hours from lunar takeoff to redocking with the CSM. Per procedure, the two LM astronauts used a small vacuum to clean up the lunar dust that had collected inside their cabin—NASA wanted as little of it as possible to be dragged into the CSM. Once that was completed, Mike Collins opened up the hatch between the two vessels and greeted his two returning crewmates.

Miscellaneous gear, boxes of lunar rocks, and two tired astronauts had to be transported through the connection tunnel. Once that was completed, Collins released the LM to orbit freely, and empty, around the Moon. Aldrin—who had been awake for most of the past three days—was exhausted, yet there was still one last important duty: the TEI: Trans-Earth Injection. They needed to fire their CSM engine one more time to break free of lunar orbit and return to Earth. Like most everything on the CSM, this was Collins's responsibility, and he had only one chance to do it right. As with Apollo 8, the burn had to be accomplished on the far side of the Moon, out of radio contact with Mission Control. Aldrin describes the moment in his book, *Magnificent Desolation*:

> Columbia's engines flared and ignited . . . right on the mark. Twenty minutes later we emerged from the back side of the Moon for the last time . . . I leaned back and closed my eyes. We were on our way home.[56]

As with Yuri Gagarin eight years before, the United States sent its three famous astronauts on a whirlwind international tour. It began with three parades in three U.S. cities—New York, Chicago, and Los Angeles—a state dinner at the White House, followed by visits to twenty-five different countries. The tour ended in Washington, DC, where all three astronauts spoke before a joint session of Congress, to whom they gifted two American flags that they had carried on their lunar mission.

In July 1969, I was fifteen years old and getting ready to start my junior year at Chaminade Preparatory High School in Canoga Park,[57] California, not far from where the mighty F-1 engine had been built and tested. My family and I were watching the news right after the re-entry, splashdown, and recovery of the Apollo 11 capsule. Armstrong, Aldrin, and Collins were ensconced in an Airstream travel trailer that had been converted into postmission living quarters. In the news broadcast, President Nixon stood outside the trailer, happily talking to the astronauts via a mic and speaker system that had been set up. The newscaster explained that the astronauts were in a two-week isolation—a voluntary quarantine—as a precaution in case they had brought back any nasty microbes or viruses from the Moon.

I mentioned how great an idea it was to take such a precaution, a comment that made my rocket engineer parents laugh.

"Why is that funny?" I asked.

My mother shook her head. "It's all for show."

Yes," agreed my father. "This so-called quarantine is silly. There's no way any living thing could survive on the Moon."

"Between the extreme temperatures of night and daylight, the intense radiation, the scarcity of water, and a complete lack of oxygen, natural life on the Moon is impossible," added my mother. "This quarantine thing is all just a big circus."

They could tell by the look on my face that I was skeptical.

"Think of it this way," said my father. "When the astronauts exited their capsule, the scuba divers and other military personnel that assisted the astronauts were exposed to them, if only briefly. So were the airmen on the helicopter, the sailors on the ship, and the doctors that examined them on that ship. And the doctors who now go in and out of that trailer to examine them—they're exposed as well. If this were a true quarantine, all of those exposures would have been avoided."

"NASA is more interested in avoiding panic," added my mother. "They're concerned that people will panic if they think the astronauts are bringing alien pathogens back with them to Earth, like some Hollywood movie. What they're doing is play-acting to make people feel better."

Living with my parents was like living with a couple of Spocks, and their logic was hard to argue with—the Moon was certainly not hospitable to any form of life we Earthlings were familiar with. Was NASA just putting on a big show to make people *feel* safer rather than *be* safer? After the exhilaration of the Moon landing, it was an unpleasant thought. Still, I was thankful to have had such intelligent parents.

A review of NASA and Apollo history, performed in preparation for this book, revealed that NASA took the possibility of alien pathogens more seriously than my parents were giving them credit for. When the astronauts splashed down, for example, the navy frogmen who greeted them were wearing specially sealed diving suits. At the opening of the hatch they threw in three quarantine suits for the astronauts to change in to. NASA did about as well as it could to prevent a remote, but possible, infection.

Even so, after two more lunar missions NASA came around to agreeing with my parents and completely abandoned the quarantine procedure. Their scientists and medical professionals had concluded that no viruses or any form of life were being brought back from the Moon.[58]

30

EPITAPHS FOR THE PIONEERS

The heavens declare the glory of God, and the firmament showeth His handiwork.

—Psalms 19:1

In 1913, at the age of thirty-two, Robert Goddard became seriously ill with tuberculosis. He eventually recovered, but the disease weakened his lungs. He passed away in 1945 shortly after being diagnosed with throat cancer. Goddard's work secured 214 patents, 131 of which were awarded after his death. NASA's Goddard Space Flight Center in Maryland is named in his honor, there is a crater on the Moon with his name on it, and a shuttle craft in the TV series *Star Trek: Next Generation* is named Goddard.

Soon after setting up shop in Fort Bliss, Wernher von Braun assigned Dieter Huzel to act as liaison between the army missile team and North American Aviation, which was getting into the liquid fuel rocket business. This position resulted in NAA hiring away Huzel from the army, and he settled into a life of perpetual springlike weather in Woodland Hills, California. He was a contributor to all the major rocket engines that poured forth from NAA-founded Rocketdyne, and published several books on rocket science. Dieter Huzel passed away on November 2, 1994.

Arthur Rudolph's contributions to history's greatest adventure were second only to those of Wernher von Braun. And yet, despite his numerous accomplishments and accolades, Arthur Rudolph's Nazi past would not be as easily whitewashed as those of his colleagues. In 1979, two years after the death of Wernher von Braun, the Office of Special Investigations[1] decided to take a closer look at Arthur Rudolph. Based on his use of slave labor at Mittelwerk during the war, the OSI strong-armed Rudolph into renouncing his U.S. citizenship and returning to Germany.

The irony of this event is that most of the German rocket scientist émigrés had, to one extent or another, some involvement in the use of slave labor during the war, especially von Braun. Compared to the treatment of the other German scientists, it appears Rudolph was scapegoated by the OSI. Had von Braun still been alive, it is possible his political influence and public celebrity may have assisted Rudolph in remaining in America—he always played the role of guardian angel for his German colleagues. But by 1979 he was gone, and so was his protection.

In 1961 Kohei Hanami, the Japanese warship commander who had initiated the fall of an unlikely string of historical dominoes by ordering the ramming of Lieutenant John Kennedy's PT boat, attended Kennedy's Washington inauguration.[2] Ironically, Hanami had also entered politics after World War II, becoming the mayor of a small village in Fukushima Prefecture, about one hundred miles north of Tokyo.[3] The date of his death does not appear to be a matter of public record. There would never have been a Moon landing if there had not been a President John F. Kennedy. There would never have been a President Kennedy if there had not been the heroism of PT-109. There would never have been the heroism of PT-109 if Captain Hanami had not decided to turn his ship and ram the small boat. To what do we owe the flying of the stars and stripes on the Sea of Tranquility? It is not so far-fetched to say we owe it to a little-known Japanese navy captain who was just trying to get home safe on a moonless Pacific battle zone night, August 2, 1943.

Sergei Korolev, arguably the greatest space pioneer of all time, did not live to see the first manned Moon landing. He had wanted his own country to be the first to land there, but in his zeal to always be the first at everything he had neglected his own health, ignoring warning signs and frequently cancelling or postponing doctor visits. On January 14, 1966, Korolev gave in to his physicians and advisors, agreeing to undergo a routine intestinal surgical procedure that he had been postponing. Unfortunately, the surgeons were unaware of how fragile Korolev's body had become from overwork and years spent in the Soviet gulags under Lenin. Internal bleeding and two abdominal tumors made the surgery overly challenging, and Korolev died on the operating table from a heart attack.[4] His close friend and colleague, cosmonaut Yuri Gagarin, was angered by what he felt was a preventable death. Still holding faith that the Soviet Union would soon land men on the Moon, Gagarin swore that he would one day scatter Korolev's ashes on the lunar surface.[5]

On April 12, 1961, cosmonaut Gagarin had become what all the Mercury astronauts had wanted so much to be: not only the first man in space, but the first to orbit the Earth. Though they may have envied his accomplishments, the American astronauts would not have envied most of what happened after. Though Gagarin was awarded the coveted Hero of the Soviet Union medal, his fame and political value started to work against him. Gagarin became such a valuable asset to the Soviet public relations machine that he was given fewer and fewer opportunities to engage in his favorite pastime: piloting fighter jet aircraft. The fear that the Soviet Union might lose its greatest propaganda symbol resulted in Gagarin being assigned to less dangerous duties. When his best friend, cosmonaut Vladimir Komarov, died in a fiery re-entry crash of a highly flawed Soyuz capsule, Gagarin became so depressed, he was permanently removed from the cosmonaut flight roster. Yuri Gagarin would never fly in space again.[6]

Like Korolev and Komarov, Gagarin would not live to see the Apollo 11 astronauts land on the Moon. On March 27, 1968, he was on a routine flight in a MiG-15 UTI, a two-man Soviet trainer jet. The craft crashed in a forested area about sixty miles northeast of Moscow. The jet buried itself so far into the frozen earth that when the first responders arrived, only a few scattered pieces of metal were visible around a large crater.[7] The exact cause of the crash was never determined. Today, every April 12 is celebrated in many countries as Yuri's Night to commemorate Gagarin's pioneering orbital flight. The two medals left on the surface of the Moon by Aldrin and Armstrong for Komarov and Gagarin will remain there, if left undisturbed, for millions of years.

With the exception of Gus Grissom, all the original Mercury astronauts lived to see retirement. After a number of years nursing political aspirations, John Glenn ran for an open Senate seat in 1974 and won. The first American to orbit the Earth became the oldest human to fly in space when he crewed the Space Shuttle in 1998. Glenn's well-known penchant for clean living rewarded him with a long and successful life. He outlived every member of the Mercury 7 astronauts, passing away in 2016 at the age of ninety-five.

After overseeing the launch of Apollo 17 in his last role as a flight director, Gene Kranz continued to work for NASA in an administrative capacity until his retirement in 1994. He flies an aerobatic airplane and is currently working on restoring a Boeing B-17 Flying Fortress. The "Kranz Dictum" that NASA must always be "tough and competent" still forms part of NASA culture to this day.

Mike Collins retired from NASA soon after the Apollo 11 mission. He took a one-year position with the Department of State as assistant secretary of state for public affairs, after which he became the director of the National Air and Space Museum in Washington, DC. He also worked for the Smithsonian

Institution and LTV Aerospace. He is the author of several books, including his 1974 autobiography, *Carrying the Fire: An Astronaut's Journeys.* Collins is also a watercolor artist, but refuses to sign his paintings on the grounds that the signature would make them artificially more valuable. He currently lives in Marco Island, Florida.

As of this writing, Edwin "Buzz" Aldrin remains active in the aerospace world. After returning to Earth from the Apollo 11 flight he was awarded the Distinguished Service Medal from the U.S. Air Force, among many other accolades. In 1988, he legally changed his first name to Buzz. A documentary about the Moon landing, *Magnificent Desolation*, detailed his numerous bouts over the years with alcoholism and depression. Today Aldrin stumps tirelessly for a future manned mission to Mars. He currently lives in Satellite Beach, Florida.

After returning to Earth as one of the most famous men in history, Neil Armstrong spent a year as the deputy associate administrator for the Office of Advanced Research and Technology. In 1971 he resigned from NASA and moved into academia as a professor of aeronautics at the University of Cincinnati. He taught there for eight years before leaving without explanation. After the Challenger disaster, on whose investigative board he served, he accepted a position on the board of Morton-Thiokol, the company who built the solid boosters blamed for the Challenger explosion.

On August 7, 2012, Armstrong underwent heart bypass surgery, which at first appeared successful. However, he soon developed complications and died in the hospital eighteen days later on August 25. Neil Armstrong had lived a life of humility, always waving off numerous offers that would have preserved his public profile, such as running for public office. To this day his surviving family, friends, and colleagues refer to him as "America's reluctant hero."

In July 1972, Ed Uhl, the young CEO of Fairchild Corporation and an acquaintance of Wernher von Braun since his White Sands days, announced that Fairchild had managed to recruit the German scientist away from NASA. The news electrified the company, which at that time had virtually no footprint in the space hardware business.[8] For von Braun, the move to the private sector after forty years of collecting government paychecks had several advantages, not the least of which was a 600 percent increase in salary. He wife was happy for another reason—the Fairchild job would allow the von Brauns to remain in their new home in Alexandria, Virginia.

Like Sergei Korolev, Wernher von Braun had been so driven, so obsessed with achieving glory in space, that he had made his personal health a low priority. In August 1975, Wernher took his family on a vacation to Canada's North Bay wilderness. While there he experienced some rectal bleeding. As before,

Wernher shrugged it off and saw no reason to see a doctor. A few weeks later while on an Alaskan hunting trip with Ed Uhl, the bleeding became much worse, and Uhl refused to allow von Braun to postpone medical attention. They flew home to Alexandria, and Uhl made certain von Braun immediately saw a specialist at Johns Hopkins Medical Center. The specialist ordered surgery without delay. Like Korolev, von Braun's body had been in declining health for some time, though he refused to acknowledge it.

The boy whose singular lifelong obsession was to pilot a rocket to the Moon would not achieve his dream. On June 16, 1977, Wernher von Braun passed away from complications related to pancreatic cancer. To avoid a media circus, his death was kept secret, and he was buried the very same day in a private ceremony in Ivy Hill Cemetery, Alexandria, Virginia. The epitaph on his tombstone is von Braun's favorite quote from Psalms: "The heavens declare the glory of God, and the firmament showeth His handiwork."

ACKNOWLEDGMENTS

I was born into a very unusual household. My father, G. Richard Morgan, and my mother, Mary Sherman Morgan, were pioneers in the post–World War II world of big rockets. How many people can say they are the offspring of not one, but two, rocket scientists? It was inevitable, therefore, that I would grow up in a home where the tenets of science were greatly valued, often discussed, and frequently practiced. Were it not for that background, it is unlikely I would have had either the skills or the disposition to write this book. So, thank you, Mom and Dad, for instilling in me and my siblings at an early age an interest in the sciences. You gave us the gifts of knowledge and wisdom, and a thirst for more of both.

Apollo 11, the first Moon landing mission, was launched on July 16, 1969. The final Moon landing mission, Apollo 17, splashed down on December 19, 1972. That means that between July 16, 2019, and December 19, 2022, there will be a number of significant golden anniversaries for NASA, the United States, and planet Earth to celebrate. Call it the Golden Age of golden anniversaries. As I write this final page for the book, that first big July 2019 anniversary has arrived. I have been aware for several years that these dates were coming up, and I occasionally toyed with the idea of writing a book to commemorate those achievements. A big thank you to my agent, Deborah Ritchken at Marsal Lyon Literary Agency, who found a way to make that happen.

Every writer needs someone to give an honest and unbiased review of his or her work, and for me that has been my good friend, Duane Ashby. He read an early version of the manuscript and passed along a number of ideas and suggestions, most of which I incorporated into the final work.

There are many engineers I was able to interview for my mother's memoir, *Rocket Girl*, whose input has been given a second life here. I had sit-downs with numerous former NAA and Rocketdyne engineers including Bill Webber, Dan Ruttle, Bill Vietinghoff, Walter Unterberg, Bill Wagner, and Irving Kanarek. I was lucky enough to have my father snag me a rare interview with his former coworker, Dieter Huzel, one of Wernher von Braun's closest associates. That was in 1988, when writing about the rocket business was just a gleam in my eye. Somehow, I managed to save the notes from that interview all these years, and was finally able to put them to use.

One last word about my father. In 2013 I stopped by his home for a visit. He was still living in Canoga Park, just a few miles from where he had spent decades working for NAA and Rocketdyne. My mother had passed away in 2004, and he was living alone. During the course of our conversation he gave me a file folder containing the first few chapters of a book he was writing. He had a great deal of firsthand knowledge about the Space Race and the Apollo program, and he wanted desperately to put his experience and knowledge down on paper and publish a book. He asked me to read over what he had written and give him some feedback. I promised I would, but warned him that writing a book requires a great deal of time and energy. I didn't mention it, but his advancing age and failing health made me doubt he would be able to complete such a project. He seemed to read my mind, waving off my concerns. "I'm pretty sure I've got a good ten years left in me," he insisted. G. Richard Morgan, Caltech grad, mechanical engineer, rocket scientist, husband, and father, passed away a year later.

This is the book he would love to have written.

NOTES

CHAPTER 1

1. Robert Goddard, high school valedictorian speech, June 1904. Quoted in Mildred K. Lehman and Milton Lehman, "Robert Goddard," *Encyclopedia Britannica*, October 1, 2019, https://www.britannica.com/biography/Robert-Goddard.
2. Asif Siddiqi, "Russia's Long Love Affair with Space," *Air & Space Magazine*, August 2007, https://www.airspacemag.com/space/russias-long-love-affair-with-space-19739095/.

CHAPTER 2

1. Bob Ward, *Dr. Space: The Life of Wernher von Braun* (Annapolis, MD: Naval Institute Press, 2005), 47.
2. Ward, *Dr. Space*, 53.

CHAPTER 3

1. The Butterfly Effect refers to the idea that small events can have nonlinear impacts on a complex system.
2. William Doyle, *PT-109: An American Epic of War, Survival, and the Destiny of John F. Kennedy* (New York: William Morrow, 2015), 97.
3. Doyle, *PT-109*, 107.

CHAPTER 4

1. Michael J. Neufeld, *Von Braun: Dreamer of Space, Engineer of War* (New York: Vintage Books, 2007), 202.
2. Neufeld, *Von Braun*, 202.
3. Neufeld, *Von Braun*, 197.
4. Neufeld, *Von Braun*, 199.
5. Neufeld, *Von Braun*, 200.
6. Neufeld, *Von Braun*, 201.

CHAPTER 5

1. Pam Rogers, "A Member of the Old Team Looks Back," *Redstone Rocket*, February 5, 1986.
2. Winston S. Churchill, *Triumph and Tragedy* (Boston: Houghton Mifflin, 1953).
3. Michael J. Neufeld, *Von Braun: Dreamer of Space, Engineer of War* (New York: Vintage Books, 2007), 212.
4. Bob Ward, *Dr. Space: The Life of Wernher von Braun* (Annapolis, MD: Naval Institute Press, 2005), 63.
5. Neufeld, *Von Braun*, 214.
6. Ward, *Dr. Space*, 63.
7. Ward, *Dr. Space*, 59.
8. Ward, *Dr. Space*, 66.
9. Huzel, interview with the author, November 1988.
10. Ward, *Dr. Space*, 66.

CHAPTER 6

1. Jay Barbree, "For Neil Armstrong, It Was a Muddy Boot in Korea before a Step on the Moon," HistoryNet, accessed July 20, 2017, http://www.historynet.com/a-wing-and-a-prayer.htm.
2. Bob Ward, *Dr. Space: The Life of Wernher von Braun* (Annapolis, MD: Naval Institute Press, 2005), 74.
3. Ward, *Dr. Space*, 74.
4. Ward, *Dr. Space*, 76.
5. Wernher von Braun, "Man on the Moon: The Journey," *Colliers*, October 18, 1952.

6. Ward, *Dr. Space*, 78.
7. Ward, *Dr. Space*, 91.

CHAPTER 7

1. Matthew Brzezinski, *Red Moon Rising: Sputnik and the Hidden Rivalries That Ignited the Space Age* (New York: Times Books, 2007), 99.
2. Korolev had no way of knowing it at the time, but the design of the R-7 turned out to be so reliable and efficient that versions of it would still be in use more than six decades later.

CHAPTER 8

1. The Saturn V, which sent three men to the moon in 1969, was only thirty-three feet in width.
2. Michael J. Neufeld, *Von Braun: Dreamer of Space, Engineer of War* (New York: Vintage Books, 2007), 269.
3. Neufeld, *Von Braun*, 220.
4. Since their inception, the Atlas and the Russian R-7 have had more than seven hundred successful launches, making the strap-on booster design synonymous with reliability.
5. Neufeld, *Von Braun*, 283.
6. Hypergolic propellants are those that ignite simply by coming in contact with each other.
7. Konstantin Gerchik, *Nezabyvayemyy Baykonur* (Moscow: Tekhnika molodezhi, 1998), 72.
8. Deborah Cadbury, *Space Race: The Epic Battle between America and the Soviet Union for Dominion in Space* (New York: Harper Collins, 2006), 214.
9. Alan Brinkley, *John F. Kennedy* (New York: Henry Holt and Company, 2012), 26.

CHAPTER 9

1. Bill Webber, interview with the author, August 5, 2002.
2. Sunny Tsiao, *Read You Loud and Clear: The Story of NASA's Spaceflight Tracking and Data Network* (NASA History Series; Washington, DC: NASA, 2008), 94.
3. Full disclosure: Mary Sherman Morgan is my mother.
4. Mark Wade, "LOX/Hydyne," Astronautix, accessed March 1, 2017, http://www.astronautix.com/l/loxhydyne.html.

CHAPTER 10

1. Betsy Kuhn, *The Race for Space* (Minneapolis: Twenty-First Century Books, 2007), 7.

2. Douglas J. Mudgway, *William H. Pickering: America's Deep Space Pioneer* (Washington, DC: Progressive Management Publications, 2008), 56.

3. Paul Dickson, *Sputnik: The Launch of the Space Race* (Toronto: Macfarlane Walter & Ross, 2001), 63.

4. Michael J. Neufeld, *Von Braun: Dreamer of Space, Engineer of War* (New York: Vintage Books, 2007), 311.

5. Gerard J. Degroot, *Dark Side of the Moon: The Magnificent Madness of the American Lunar Quest* (New York: New York University Press, 2006), 88.

6. Dickson, *Sputnik*, 88.

7. Degroot, *Dark Side of the Moon*, 90.

8. Degroot, *Dark Side of the Moon*, 123.

CHAPTER 11

1. Deborah Cadbury, *Space Race: The Epic Battle between America and the Soviet Union for Dominion in Space* (New York: Harper Collins, 2006), 191.

2. When a later Pioneer probe was successful, the air force gave it the name Pioneer 1 and renamed the original failed probe Pioneer 0.

3. Cadbury, *Space Race*, 188.

4. Cadbury, *Space Race*, 188.

5. An "impactor" probe is one designed to crash into its target rather than attempt a soft landing.

CHAPTER 12

1. Wally Schirra, *Schirra's Space* (Annapolis, MD: Naval Institute Press, 1988), 46.

2. Schirra, *Schirra's Space*, 45.

3. Schirra, *Schirra's Space*, 54.

4. Deborah Cadbury, *Space Race: The Epic Battle between America and the Soviet Union for Dominion in Space* (New York: Harper Collins, 2006), 215.

5. It was this combination of gumption and gusto that would later cause Tom Wolfe to entitle his 1979 astronaut memoir *The Right Stuff*.

6. It is one of the many ironies of American space history that Glennan, who spent much of his career in the film business, would later be portrayed on screen by an actor in the film version of *The Right Stuff*. John P. Ryan turned in a terrific performance

as Glennan, though the credits identify him as "Head of the Program" rather than as Thomas Keith Glennan.

7. NASA Content Administrator, "Mercury Project Overview–Astronaut Selection," National Aeronautics and Space Administration, November 30, 2006, https://www.nasa.gov/mission_pages/mercury/missions/astronaut.html.

8. Tom Wolfe, *The Right Stuff* (New York: Bantam Books, 2001 [originally published in 1979]), 90.

9. Wolfe, *The Right Stuff*, 90.

10. Michael J. Neufeld, *Von Braun: Dreamer of Space, Engineer of War* (New York: Vintage Books, 2007), 339.

11. Neufeld, *Von Braun*, 339.

12. Dieter Huzel would be assigned to the same North American research group that employed Mary Sherman Morgan, America's first female rocket scientist, and the mother of this author.

13. Neufeld, *Von Braun*, 338.

14. Though usually referred to as a spacesuit, the actual technical term is pressure suit.

15. Schirra, *Schirra's Space*, 139.

16. In 1974 John Glenn would capitalize on the recently opened Walt Disney World by re-entering the central Florida hotel business, a move that would prove highly successful.

17. Schirra, *Schirra's Space*, 140.

18. Homer H. Hickam, Jr., *The Rocket Boys* (New York: Delacorte Press, 1998), 336.

CHAPTER 13

1. The precursor to NASA.

2. Gene Kranz, *Failure Is Not an Option: Mission Control from Mercury to Apollo 13 and Beyond* (New York: Simon & Schuster, 2009), 68.

3. Michael J. Neufeld, *Von Braun: Dreamer of Space, Engineer of War* (New York: Vintage Books, 2007), 337.

4. The film version of *The Right Stuff* was directed by Philip Kaufman and released in 1983 by Warner Bros.

5. Chris Kraft, *Flight: My Life in Mission Control* (New York: Dutton, 2001), 83.

6. Kranz, *Failure Is Not an Option*, 22.

7. Unless you count Homer Hickam and about a hundred Germans.

8. Later renamed Johnson Space Center in 1973.

CHAPTER 14

1. Members of Carrier Air Group 19 were in the process of training for combat in the Korean War, but the war ended before any of the men saw combat.

2. Colin Burgess, *Freedom 7: The Historic Flight of Alan B. Shepard, Jr.* (New York: Springer-Praxis, 2014), 12.

3. Wally Schirra, *Schirra's Space* (Annapolis, MD: Naval Institute Press, 1988), 71.

4. Schirra, *Schirra's Space*, 72.

5. Gene Kranz, *Failure Is Not an Option: Mission Control from Mercury to Apollo 13 and Beyond* (New York: Simon & Schuster, 2009), 25.

6. Rod Pyle, *Heroes of the Space Age* (New York: Prometheus Books, 2019), 24.

7. James Donovan, *Shoot for the Moon: The Space Race and the Extraordinary Voyage of Apollo 11* (New York: Little, Brown, and Company, 2019), 85.

8. Kranz, *Failure Is Not an Option*, 52.

9. Donovan, *Shoot for the Moon*, 85.

10. Kranz, *Failure Is Not an Option*, 53.

11. NASA would later discover this action was less risky than they thought; the 100 percent oxygen atmosphere within the capsule quickly evaporated the urine in the suit.

12. "Flight" was the title given to their manager, Chris Kraft.

CHAPTER 15

1. In comparison, the Redstone A-7 engine that put the first Mercury 7 astronauts into space had a maximum thrust of only seventy-eight thousand pounds.

2. Bob Ward, *Dr. Space: The Life of Wernher von Braun* (Annapolis, MD: Naval Institute Press, 2005), 136.

3. I catalogued my mother's life in the book *Rocket Girl: The Story of Mary Sherman Morgan, America's First Female Rocket Scientist* (Buffalo, NY: Prometheus Books, 2013).

4. Andrew Young, *The Saturn V F-1 Engine: Powering Apollo into History* (Chichester, UK: Praxis Publishing, 2009), 47.

5. William E. Anderson and Viggo Yang, *Liquid Rocket Engine Combustion Instability* (Reston, VA: American Institute of Aeronautics and Astronautics, 1995), https://arc.aiaa.org/doi/book/10.2514/4.866371.

6. Young, *The Saturn V F-1 Engine*, 58.

7. Improved safety and further redundancy also worked into the decision.

8. As of this writing, there is an F-1 on outdoor public display in the parking lot of the recently merged Aerojet-Rocketdyne in Canoga Park, California.

CHAPTER 16

1. *Wikipedia*, Gemini (constellation), last modified April 12, 2019, https://en.wikipedia.org/wiki/Gemini.

2. Ben Evans, "A Bad Call: The Accident Which Almost Lost Project Gemini," AmericaSpace, March 5, 2012, http://www.americaspace.com/2012/03/05/a-bad-call-the-accident-which-almost-lost-project-gemini/.

3. "Murphy's Law": A philosophical view of the world that says anything that can go wrong, will go wrong.

4. Gene Kranz, *Failure Is Not an Option: Mission Control from Mercury to Apollo 13 and Beyond* (New York: Simon & Schuster, 2009), 126.

5. Kranz, *Failure Is Not an Option*, 130.

6. Kranz, *Failure Is Not an Option*, 130.

7. *The Unsinkable Molly Brown* was a popular Broadway musical at the time.

CHAPTER 17

1. By the time the Space Shuttle came along there would be a total of thirty-three teams, with color assignments as exotic as granite, garnet, midnight, and crystal. When they ran out of colors, they used the names of stars, heavenly objects, and anything connected to the sky—like, for example, Kitty Hawk.

2. Gene Kranz, *Failure Is Not an Option: Mission Control from Mercury to Apollo 13 and Beyond* (New York: Simon & Schuster, 2009), 134.

3. The actual mission elapsed time was seven days, twenty-two hours, fifty-five minutes.

CHAPTER 18

1. Michael Collins, *Carrying the Fire: An Astronaut's Journeys* (New York: Farrar, Straus, and Giroux, 2009), 154.

2. Collins, *Carrying the Fire*, 154.

3. Dan Kendall, "Apollo 7 and the Importance of Guenter Wendt," National Space Centre, October 10, 2018, https://spacecentre.co.uk/blog-post/apollo-7-guenter-wendt/.

4. Wally Schirra, *Schirra's Space* (Annapolis, MD: Naval Institute Press, 1988), 157.

5. Rod Pyle, *Heroes of the Space Age* (New York: Prometheus Books, 2019), 124.

6. A name given by the astronauts, due to its spherical shape.

7. Gemini VIII Voice Communications, NASA, March 1966, 75.

8. Jim Dumoulin, "Project Gemini XI," NASA, last updated August 25, 2000, https://science.ksc.nasa.gov/history/gemini/gemini-xi/gemini-xi.html.

9. Gene Kranz, *Failure Is Not an Option: Mission Control from Mercury to Apollo 13 and Beyond* (New York: Simon & Schuster, 2009), 90.

CHAPTER 19

1. Michael J. Neufeld, *Von Braun: Dreamer of Space, Engineer of War* (New York: Vintage Books, 2007), 356.

2. James R. Hansen, "The Rendezvous That Was Almost Missed," NASA Langley Research Center, December 1, 1992, https://www.nasa.gov/centers/langley/news/fact sheets/Rendezvous.html.

3. Deborah Cadbury, *Space Race: The Epic Battle between America and the Soviet Union for Dominion in Space* (New York: Harper Collins, 2006), 263.

4. Cadbury, *Space Race*, 264.

5. Although everyone involved continued to refer to it using the craft's original pronunciation, "lem."

6. The design evolved from a craft with five legs, to four legs, to three legs, then back to four legs.

7. "Six Stories from Developing the Lunar Module," Smithsonian National Air and Space Museum, August 13, 2016, https://airandspace.si.edu/stories/editorial/six-stories-developing-lunar-module.

8. The final cost would balloon to $2 billion.

9. Kevin M. Rusnak, "Interview with Thomas J. Kelly," NASA Johnson Space Center Oral History Portal, September 19, 2000, https://historycollection.jsc.nasa.gov/JSCHistoryPortal/history/oral_histories/KellyTJ/KellyTJ_9-19-00.htm.

10. My father, George Richard Morgan, worked for Marquardt for several years and participated in the initial designs of the LM reaction control system. By the mid-1960s, however, he had moved across the San Fernando Valley to Rocketdyne, preferring to work on the "big engines."

11. Rusnak, "Interview with Thomas J. Kelly."

12. Due to their mixture ratio and differences in propellant densities, one tank would be much larger than the other, hence unsymmetrical.

13. Rusnak, "Interview with Thomas J. Kelly."

14. Gene Kranz, *Failure Is Not an Option: Mission Control from Mercury to Apollo 13 and Beyond* (New York: Simon & Schuster, 2009), 217.

15. Kranz, *Failure Is Not an Option*, 220.

16. One year before his death in 2002, Kelly published a memoir of how the LM was designed and built. The book is entitled *Moon Lander: How We Developed the Apollo Lunar Module* (Washington, DC: Smithsonian Books, 2001).

CHAPTER 20

1. Bob Ward, *Dr. Space: The Life of Wernher von Braun* (Annapolis, MD: Naval Institute Press, 2005), 132.

2. Ward, *Dr. Space*, 132.
3. Ward, *Dr. Space*, 132.
4. Ward, *Dr. Space*, 134.
5. William E. Burrows, *The New Ocean: The Story of the First Space Age* (New York: Random House, 1998), 332.
6. John M. Logsdon, "John F. Kennedy and NASA," NASA: History, May 22, 2015, https://www.nasa.gov/feature/john-f-kennedy-and-nasa.

CHAPTER 21

1. Douglas Mudgway, *William H. Pickering: America's Deep Space Pioneer* (Washington, DC: Progressive Management Publications, 2008).
2. Deborah Cadbury, *Space Race: The Epic Battle between America and the Soviet Union for Dominion in Space* (New York: Harper Collins, 2006), 262.
3. Cadbury, *Space Race*, 289.
4. Mudgway, *William H. Pickering*, 156.

CHAPTER 22

1. G. Richard Morgan, interview with the author, February 10, 2010.
2. Michael J. Neufeld, *Von Braun: Dreamer of Space, Engineer of War* (New York: Vintage Books, 2007), 398.
3. Neufeld, *Von Braun*, 398.

CHAPTER 23

1. Gene Kranz, *Failure Is Not an Option: Mission Control from Mercury to Apollo 13 and Beyond* (New York: Simon & Schuster, 2009), 197.
2. The command module was the capsule that housed the astronauts. The service module was the cylindrical section just beneath the capsule that contained propulsion systems, propellant storage, electrical power generation, and some food and water. For this book I will refer to both sections together as the CSM.
3. Kranz, *Failure Is Not an Option*, 197.
4. Kranz, *Failure Is Not an Option*, 197.
5. Kranz, *Failure Is Not an Option*, 199.
6. Apollo 204 Review Board, "Report of the Apollo 204 Review Board," NASA History Office, last updated February 23, 2014, https://history.nasa.gov/Apollo204/as204report.html.

7. According to NASA's summary of the Apollo 1 accident and its aftermath, there was evidence White had been making an attempt to open the hatch. "Summary: Apollo 1: The Fire," NASA History Office, accessed November 20, 2019, https://history.nasa.gov/SP-4029/Apollo_01a_Summary.htm.
8. Apollo 204 Review Board, "Report of the Apollo 204 Review Board."
9. Apollo 204 Review Board, "Report of the Apollo 204 Review Board."
10. Ellington Air Force Base has since been renamed the Johnson Space Center.
11. Bob Granath, "Theodore Freeman Honored in 50th Anniversary Memorial Ceremony," NASA, November 3, 2014, https://www.nasa.gov/content/theodore-freeman-honored-in-50th-anniversary-memorial-ceremony.
12. Michael Collins, *Carrying the Fire: An Astronaut's Journeys* (New York: Farrar, Straus, and Giroux, 2009), 271.
13. Apollo 204 Review Board, "Report of the Apollo 204 Review Board."
14. U.S. Senate Committee on Aeronautical and Space Sciences, "Apollo 204 Accident," January 30, 1968, NASA History Office, https://history.nasa.gov/as204_senate_956.pdf, 10.
15. U.S. Senate Committee on Aeronautical and Space Sciences, "Apollo 204 Accident," January 30, 1968, NASA History Office, https://history.nasa.gov/as204_senate_956.pdf, 10.
16. George Leopold, *Calculated Risk: The Supersonic Life and Times of Gus Grissom* (West Lafayette, IN: Purdue University Press, 2016), 198.
17. Kranz, *Failure Is Not an Option*, 203.
18. Kranz, *Failure Is Not an Option*, 203.

CHAPTER 24

1. Amy Shira-Teitel, "Vintage Space: What Happened to Apollos 2 and 3?" *Popular Science*, October 28, 2013, https://www.popsci.com/blog-network/vintage-space/what-happened-apollos-2-and-3#page-2.
2. In July 1987, Dan Ruttle, Stephen Morgan, myself, and several members of the Pacific Rocket Society launched a nitric acid/furfuryl alcohol rocket in Nevada's Smoke Creek Desert to an altitude of twenty thousand feet.
3. Orlando Bongat, "Rocket Park: Saturn V," NASA, last updated September 16, 2011, https://www.nasa.gov/centers/johnson/rocketpark/saturn_v.html.
4. Anna Heiney, "The Crawlers," NASA, last updated June 20, 2018, https://www.nasa.gov/content/the-crawlers. Today, more than a half century later, the crawler-transporter is still in service.
5. Craig Nelson, *Rocket Men: The Epic Story of the First Men on the Moon* (New York: Penguin Group, 2009), 1.
6. "Saturn V Moon Rocket: Historical Snapshot," Boeing Corporation, accessed March 31, 2019, https://www.boeing.com/history/products/saturn-v-moon-rocket.page.

7. W. David Woods, *How Apollo Flew to the Moon* (Chichester: Praxis Publishing, 2008), 28.

8. It would be another sixteen years before America would launch its first woman into space, astronaut Sally Ride.

9. Andrew Chaikin, *A Man on the Moon: The Voyages of the Apollo Astronauts* (New York: Penguin Group, 1995), 54.

10. Robert Kurson, *Rocket Men* (New York: Random House, 2018), 152.

11. Amy Shira-Teitel, "NASA's Gutsy First Launch of the Saturn V Moon Rocket," Space.com, November 15, 2012, https://www.space.com/18505-nasa-moon-rocket-saturn-v-history.html.

12. Gene Kranz, *Failure Is Not an Option: Mission Control from Mercury to Apollo 13 and Beyond* (New York: Simon & Schuster, 2009), 209.

13. Kranz, *Failure Is Not an Option*, 210.

14. Bob Granath, "Apollo 4 Was First-Ever Launch from NASA's Kennedy Space Center," NASA, November 9, 2017, https://www.nasa.gov/feature/apollo-4-was-first-ever-launch-from-nasas-kennedy-space-center.

15. Granath, "Apollo 4."

16. It would not be until years later, when the Space Shuttle was being designed, that propulsion engineers would finally find a permanent solution to the pogo effect.

17. Gemini XII, November 1966.

18. Yes, 66 percent of the crew was named Walter.

19. Ironically this division would be later resolved through consolidation as NAA and Rockwell soon merged to become North American Rockwell.

20. NASA, Apollo 7: Technical Air-To-Ground Voice Transcriptions, NASA, last updated October 24, 2008, https://history.nasa.gov/alsj/a410/AS07_TEC.PDF.

21. Schirra had effectively dodged that bullet by announcing his retirement from NASA just before the Apollo 7 launch.

CHAPTER 25

1. Jeffrey Kluger, *Apollo 8: The Thrilling Story of the First Mission to the Moon* (New York: Henry Holt and Company, 2017), 165.

2. Kluger, *Apollo 8*, 169.

3. As of the writing of this book, no country other than the United States has sent a manned mission beyond Earth orbit.

4. Kluger, *Apollo 8*, 171.

5. Kluger, *Apollo 8*, 188.

6. Kluger, *Apollo 8*, 212.

7. NASA astronauts apparently despised the term A-OK.

8. Kluger, *Apollo 8*, 235.

9. Kluger, *Apollo 8*, 236.

10. Kluger, *Apollo 8*, 245.
11. Kluger, *Apollo 8*, 245.

CHAPTER 26

1. James Donovan, *Shoot for the Moon: The Space Race and the Extraordinary Voyage of Apollo 11* (New York: Little, Brown, and Company, 2019), 278.
2. Donovan, *Shoot for the Moon*, 263.
3. Michael Collins, *Carrying the Fire: An Astronaut's Journeys* (New York: Farrar, Straus, and Giroux, 2009), 288.
4. Collins, *Carrying the Fire*, 288.
5. Collins, *Carrying the Fire*, 314.

CHAPTER 27

1. James Donovan, *Shoot for the Moon: The Space Race and the Extraordinary Voyage of Apollo 11* (New York: Little, Brown, and Company, 2019), 283.
2. Despite performing admirably, Apollo 9 would be Schweickart's only spaceflight.
3. NASA Content Administrator, "Apollo 9: Spider's First Mission," NASA, January 9, 2018, https://www.nasa.gov/mission_pages/apollo/missions/apollo9.html.
4. Denise Chow, "Mystery of Moon's Lumpy Gravity Explained," Space.com, May 30, 2013, https://www.space.com/21364-moon-gravity-mascons-mystery.html.
5. NASA Content Administrator, "Apollo 10," NASA, July 8, 2009, https://www.nasa.gov/mission_pages/apollo/missions/apollo10.html.

CHAPTER 28

1. Jason Catanzariti, "Flight Training for Apollo: An Interview with Astronaut Harrison Schmitt," *The Space Review*, December 10, 2012, http://www.thespacereview.com/article/2199/1.
2. It was also sometimes referred to as the LLTV: Lunar Landing Training Vehicle.
3. Betsy Mason, "The Incredible Things NASA Did to Train Apollo Astronauts," *Wired*, July 20, 2011, https://www.wired.com/2011/07/moon-landing-gallery/.
4. Michael Collins, *Carrying the Fire: An Astronaut's Journeys* (New York: Farrar, Straus, and Giroux, 2009), 259.
5. Mason, "The Incredible Things NASA Did."
6. "Armstrong's Close Call," *Smithsonian Air & Space Magazine*, May 13, 2009, https://www.airspacemag.com/videos/armstrongs-close-call/.

7. Of the five LLRVs that were built, three would be destroyed in crashes.

8. Collins, *Carrying the Fire*, 256.

9. Collins, *Carrying the Fire*, 328.

10. Collins, *Carrying the Fire*, 329.

11. Collins, *Carrying the Fire*, 329.

12. Gene Kranz, *Failure Is Not an Option: Mission Control from Mercury to Apollo 13 and Beyond* (New York: Simon & Schuster, 2009), 234.

13. Kranz, *Failure Is Not an Option*, 263.

CHAPTER 29

1. Margy Bloom, "Pregnant Guppy: The Plane That Won the Space Race," *Pilotmag*, May/June 2010, https://issuu.com/pilotmag/docs/mayjune2010.

2. Wernher had become such a worldwide celebrity that he and his family took the trip under assumed names.

3. Michael J. Neufeld, *Von Braun: Dreamer of Space, Engineer of War* (New York: Vintage Books, 2007), 430.

4. Neufeld, *Von Braun*, 433.

5. Gene Kranz, *Failure Is Not an Option: Mission Control from Mercury to Apollo 13 and Beyond* (New York: Simon & Schuster, 2009), 274.

6. Michael Collins, *Carrying the Fire: An Astronaut's Journeys* (New York: Farrar, Straus, and Giroux, 2009), 364.

7. Kranz, *Failure Is Not an Option*, 274.

8. Buzz Aldrin, with Ken Abraham, *Magnificent Desolation: The Long Journey Home from the Moon* (New York: Harmony Books, 2009), 7.

9. Collins, *Carrying the Fire*, 374.

10. Jeffrey Kluger, *Apollo 8: The Thrilling Story of the First Mission to the Moon* (New York: Henry Holt and Company, 2017), 207.

11. "Apollo Flight Journal: Day 4, Part 1, Entering Lunar Orbit," last updated February 10, 2017, NASA History Office, https://history.nasa.gov/afj/ap11fj/11day4-loi1.html.

12. Collins, *Carrying the Fire*, 391.

13. Kranz, *Failure Is Not an Option*, 277.

14. Kranz, *Failure Is Not an Option*, 277.

15. Neil Armstrong, Michael Collins, and Edwin E. Aldrin, *First on the Moon* (New York: Little, Brown, and Company, 1970), 349.

16. Collins, *Carrying the Fire*, 397.

17. The joke being that in a weightless space environment, there really is no "upside down."

18. James Donovan, *Shoot for the Moon: The Space Race and the Extraordinary Voyage of Apollo 11* (New York: Little, Brown, and Company, 2019), 355.

19. Catherine Thimmesh, *Team Moon: How 400,000 People Landed Apollo 11 on the Moon* (New York: Houghton Mifflin, 2006), 148.

20. Kranz, *Failure Is Not an Option*, 288.

21. Armstrong, Collins, and Aldrin, *First on the Moon*, 288.

22. Kevin M. Rusnak, "Oral History of Robert L. Carlton," Johnson Space Center History Project, April 10, 2001, https://historycollection.jsc.nasa.gov/JSCHistoryPortal/history/oral_histories/CarltonRL/carltonrl.htm.

23. Thimmesh, *Team Moon*. 167.

24. Kranz, *Failure Is Not an Option*, 291.

25. Aldrin, *Magnificent Desolation*, 19.

26. The LM was traveling above the lunar surface at fifty-eight feet per second—about forty miles per hour.

27. Armstrong, Collins, and Aldrin, *First on the Moon*, 289.

28. Aldrin, *Magnificent Desolation*, 20.

29. Aldrin, *Magnificent Desolation*, 21.

30. Aldrin, *Magnificent Desolation*, 21.

31. Referring to the descent engine.

32. Ten years later, Trivial Pursuit would become the most popular board game in America for a while. The first time I played this game I was given the question: "What was the first word spoken on the Moon?" I answered "Contact"—the first of the two words Aldrin spoke immediately after landing ("Contact light.") The other team shook their heads and said, "Houston." I knew I was right, but it did my team no good.

33. "Mission Transcripts: Apollo 11," Johnson Space Center, July 16, 2010, https://historycollection.jsc.nasa.gov/JSCHistoryPortal/history/mission_trans/apollo11.htm.

34. Armstrong, Collins, and Aldrin, *First on the Moon*, 292.

35. Betsy Kuhn, *The Race for Space* (Minneapolis: Twenty-First Century Books, 2007), 79.

36. Collins, *Carrying the Fire*, 402.

37. Aldrin, *Magnificent Desolation*, 43.

38. Collins, *Carrying the Fire*, 401.

39. These events were dramatized in the 2000 film *The Dish*, directed by Rob Sitch. I had the opportunity to see the U.S. premier of this film at the Sundance Film Festival. I highly recommend it.

40. John Sarkissian, "The Parkes Observatory's Support of the Apollo 11 Mission," Parkes Observatory, February 25, 2009, http://www.parkes.atnf.csiro.au/news_events/apollo11/.

41. Armstrong, Collins, and Aldrin, *First on the Moon*, 317.

42. Aldrin, *Magnificent Desolation*, 30.

43. Armstrong, Collins, and Aldrin, *First on the Moon*, 317.

44. Aldrin, *Magnificent Desolation*, 30.

45. Aldrin, *Magnificent Desolation*, 32.

46. After he returned home, Aldrin was told that Armstrong's wife Jan had made the following comment while watching Aldrin leave the LM: "Wouldn't that be something if they locked themselves out?" Armstrong, Collins, and Aldrin, *First on the Moon*, 325.

47. Aldrin, *Magnificent Desolation*, 33.

48. Aldrin, *Magnificent Desolation*, 33.

49. Andrew Chaikin, *A Man on the Moon: The Voyages of the Apollo Astronauts* (New York: Penguin Group, 1995), 214.

50. As of the writing of this book, the LLRR continues to function. Earth-bound instruments with improved sensitivity have since been built that have improved the precision of measurement to within inches.

51. Walter S. Kiefer, "Apollo 11 Mission, Science Experiments: Passive Seismic," Lunar Planetary Institute, accessed April 1, 2019, https://www.lpi.usra.edu/lunar/missions/apollo/apollo_11/experiments/pse/.

52. The Hasselblad camera used by the Apollo astronauts had no viewfinder because their helmet faceplates would prevent eyeing up to one, so the astronauts had to point and hope. Over the coming weeks, NASA would be criticized about the fact that most of the photos that included an astronaut were of Aldrin. The reason was simple: Armstrong was tasked with being the mission photographer.

53. Aldrin, *Magnificent Desolation*, 44.

54. Before leaving the LM behind for good, Aldrin would remove the breaker and keep it as a permanent memento.

55. Aldrin, *Magnificent Desolation*, 47.

56. Aldrin, *Magnificent Desolation*, 51.

57. Shortly thereafter, to increase property values, part of the town rebranded itself, changing its name to West Hills.

58. Ironically, it would turn out that the astronauts and others who came in contact with the lunar particles were in more danger than my parents and the scientific community credited, albeit not from a biological threat. In 2005 NASA announced new findings—that the inhalation of lunar dust could have a devastating effect on one's health. Inhalation of lunar dust, it was now understood, could cause a health hazard similar to silicosis. See "Don't Breathe the Moondust," NASA Science Newsletter, April 22, 2005, https://science.nasa.gov/science-news/science-at-nasa/2005/22apr_dontinhale.

CHAPTER 30

1. The OSI was a division of the U.S. Justice Department tasked with investigating war crimes, especially those from World War II.

2. The president-elect had personally invited him.

3. "Survivors and Witnesses," *EEE Reporter* (blog), August 6, 2010, http://ee reporter.blogspot.com/2010/08/mar-85-and-he-asked-them-how-many.html.

4. Piers Bizony and Jamie Doran, *Starman: The Truth behind the Legend of Yuri Gagarin* (New York: Walker Books, May 1, 2011), 185.

5. Bizony and Doran, *Starman*, 187.

6. Bizony and Doran, *Starman*, 213.

7. Bizony and Doran, *Starman*, 213.

8. Bob Ward, *Dr. Space: The Life of Wernher von Braun* (Annapolis, MD: Naval Institute Press, 2005), 204.

INDEX